Building Enterprise-Ready Azure Infrastructure

Learn Security, Scalability, and Reliability with IaC & DevOps

Roshan Gavandi

Apress®

Building Enterprise-Ready Azure Infrastructure: Learn Security, Scalability, and Reliability with IaC & DevOps

Roshan Gavandi
Mumbai, Maharashtra, India

ISBN-13 (pbk): 979-8-8688-1810-3 ISBN-13 (electronic): 979-8-8688-1811-0
https://doi.org/10.1007/979-8-8688-1811-0

Copyright © 2025 by Roshan Gavandi

This work is subject to copyright. All rights are reserved by the Publisher, whether the whole or part of the material is concerned, specifically the rights of translation, reprinting, reuse of illustrations, recitation, broadcasting, reproduction on microfilms or in any other physical way, and transmission or information storage and retrieval, electronic adaptation, computer software, or by similar or dissimilar methodology now known or hereafter developed.

Trademarked names, logos, and images may appear in this book. Rather than use a trademark symbol with every occurrence of a trademarked name, logo, or image we use the names, logos, and images only in an editorial fashion and to the benefit of the trademark owner, with no intention of infringement of the trademark.

The use in this publication of trade names, trademarks, service marks, and similar terms, even if they are not identified as such, is not to be taken as an expression of opinion as to whether or not they are subject to proprietary rights.

While the advice and information in this book are believed to be true and accurate at the date of publication, neither the authors nor the editors nor the publisher can accept any legal responsibility for any errors or omissions that may be made. The publisher makes no warranty, express or implied, with respect to the material contained herein.

Managing Director, Apress Media LLC: Welmoed Spahr
Acquisitions Editor: Aditee Mirashi
Editorial Assistant: Jacob Shmulewitz

Cover designed by eStudioCalamar

Distributed to the book trade worldwide by Springer Science+Business Media New York, 1 New York Plaza, New York, NY 10004. Phone 1-800-SPRINGER, fax (201) 348-4505, e-mail orders-ny@springer-sbm.com, or visit www.springeronline.com. Apress Media, LLC is a Delaware LLC and the sole member (owner) is Springer Science + Business Media Finance Inc (SSBM Finance Inc). SSBM Finance Inc is a **Delaware** corporation.

For information on translations, please e-mail booktranslations@springernature.com; for reprint, paperback, or audio rights, please e-mail bookpermissions@springernature.com.

Apress titles may be purchased in bulk for academic, corporate, or promotional use. eBook versions and licenses are also available for most titles. For more information, reference our Print and eBook Bulk Sales web page at http://www.apress.com/bulk-sales.

Any source code or other supplementary material referenced by the author in this book is available to readers on GitHub. For more detailed information, please visit https://www.apress.com/gp/services/source-code.

If disposing of this product, please recycle the paper

Table of Contents

About the Author xiii

About the Technical Reviewer xv

Introduction xvii

Chapter 1: Introduction to Azure Infrastructure 1
   1.1 Cloud Adoption Trends and Enterprise Use Cases 2
   1.2 Overview of Azure Core Services (Compute, Storage, Networking) 3
      Compute: The Engine Room of Azure 3
      Storage: The Backbone of Persistence 5
      Networking: The Fabric That Connects Everything 6
   1.3 Designing Scalable and Secure Infrastructure 7
      The Principle of Scalability: Planning for Growth, Not Guesswork 7
      The Principle of Security: Built-In, Not Bolted-On 9
      Design Patterns for Enterprise Resilience 10
   1.4 Logical Constructs in Enterprise Azure: Understanding Service Groups 11
   1.5 Azure Governance and Compliance 14
      Management Groups and Hierarchical Control 14
      Azure Policy: Guardrails, Not Roadblocks 14
      Role-Based Access Control (RBAC): Least Privilege at Scale 15
      Cost Management and Guardrails for Financial Control 16
      Governance for Regulated Industries 16
   1.6 Summary 17

TABLE OF CONTENTS

Chapter 2: Infrastructure as Code (IaC) in Azure 19

2.1 Introduction to IaC and Its Benefits 20
Why IaC Is Essential in Enterprise Azure Environments 20
Declarative vs. Imperative 21
Integrating with DevOps and CI/CD 22

2.2 Writing and Deploying Terraform for Azure 22
The Terraform Workflow: Plan, Apply, and Evolve 22
State Management and Remote Back Ends 24
Modules: Building Reusable Infrastructure Constructs 25
Integrating with CI/CD Pipelines 25

2.3 Automating Deployments with Bicep and ARM Templates 26
The Evolution from ARM to Bicep 27
Bicep Modules and Reuse 28
Deployment Modes and Template Specs 28
Incremental Mode: Default, Safe, and Evolutionary 29
Complete Mode: Deterministic, Declarative, and Destructive 30
Choosing the Right Mode: Strategic Guidance for Architects 31
Parameterization and Secret Management 32
CI/CD Integration and Linting 33

2.4 Best Practices for IaC in Large-Scale Environments 34
Adopt Modular Architecture: Build Infrastructure Like Software 35
Separate Configuration from Logic 36
Embed Security and Governance from Day Zero 36
Use CI/CD Pipelines for Delivery and Drift Detection 37
Manage State Securely and Centrally 37
Adopt a GitOps and Change Management Model 38

2.5 Engineering Multi-region Deployments with Safe Deployment Practices and GitOps in Azure DevOps 38
Why Multi-region IaC Is No Longer Optional 39
Matrix Strategy: Declarative Global Infrastructure with Local Customization 40
Introducing Safe Deployment Practices (SDP) in Infrastructure Pipelines 41

GitOps Flow: Commit-Driven Infrastructure Life Cycle 42

Real-World Use Case: Safe Rollout for Global FinTech Infrastructure............................. 44

2.6 Summary... 45

Chapter 3: Azure Networking and Security 47

3.1 Virtual Networks, Subnets, and Private Endpoints 48

Designing Azure Virtual Networks: Your Logical Data Center 49

Subnets and Network Segmentation.. 49

Private Endpoints: Securely Consuming PaaS Services 50

Real-World Use Case: FinTech Microservices in a Secure Mesh.. 54

3.2 Azure Firewall, DDoS Protection, and Network Security Groups (NSGs) 55

Azure Firewall: The Cloud-Native, Stateful Packet Inspection Layer 55

DDoS Protection: Absorbing Internet-Scale Attacks 56

Network Security Groups (NSGs): The Access Gatekeepers 57

Layered Security in Practice: A Government Services Portal 59

Networking Layer 61

Perimeter Security Layer 61

Monitoring Layer 61

Application and Data Layer... 62

3.3 Zero Trust Security Model in Azure 66

Principle 1: Verify Explicitly.. 67

Principle 2: Use Least Privilege Access 68

Principle 3: Assume Breach... 68

Data Security and Labeling 70

Enforcing Conditional Access with Microsoft Graph PowerShell SDK.................................... 71

3.4 Implementing Hybrid Cloud Networking 73

VPN Gateway: Encrypted Connectivity Over Public Infrastructure.. 73

ExpressRoute: Private, Dedicated Connectivity for Enterprise-Grade Performance 74

Evaluating Hybrid Connectivity: VPN Gateway vs. ExpressRoute.. 75

Azure VPN Gateway: Internet-Based Secure Connectivity 75

Azure ExpressRoute: Private, SLA-Backed Connectivity 76

Architectural Comparison Overview 76

TABLE OF CONTENTS

    Decision Criteria: When to Choose What 77
    Strategic Recommendation 78
    Networking Layer 80
    Perimeter Security Layer 80
    Monitoring Layer 81
    Application and Data Layer 81
    Extending Governance and Identity with Azure Arc 82
    Routing, DNS, and Shared Services 82
    Deploying an Azure Virtual Network Gateway with Bicep 83
  3.5 Leveraging Azure Virtual Network Manager for Consistent Network Security 85
  3.6 Summary 87

Chapter 4: High Availability and Disaster Recovery 89

  4.1 Designing for High Availability (VM Scale Sets, Load Balancers) 90
  4.2 Disaster Recovery with Azure Site Recovery (ASR) 98
    Continuous Replication and Recovery Architecture 98
    Orchestrated Failover with Recovery Plans 100
    Failback and Cost Optimization 101
    Security and Policy Integration 101
  4.3 Backup and Data Protection Strategies 102
    Azure Backup: Managed, Policy-Based Protection at Scale 103
    Point-in-Time Restore (PITR) for Azure Databases 103
    Immutable Blob Storage and Object Versioning 104
    Protecting Containerized State and AKS Workloads 105
    Governance, Monitoring, and Compliance 107
  4.4 Multi-region Deployments and Failover 108
    Deployment Topologies: Active-Passive vs. Active-Active 109
    Data Synchronization and Consistency Models 110
    Global Traffic Distribution and Routing 110
    Active-Active Multi-region Architecture with Azure Front Door and Cosmos DB ... 111
    DNS-Level Resilience with Azure Traffic Manager 112

Nested Traffic Manager with Azure Service Fabric Across Four Regions 113
How Nested Traffic Manager Works 114
Orchestration, Automation, and Recovery Readiness 115
Governance, Compliance, and Residency 116
4.5 Navigating the Complexity of On-Prem to Azure Failover Scenarios 116
4.6 Summary 118

Chapter 5: Azure Kubernetes Service (AKS) for Enterprise Workloads 121

5.1 Introduction to AKS and Kubernetes Fundamentals 122
Kubernetes Control Plane vs. Node Plane 124
Kubernetes Workload Primitives 125
Node Pools and OS Options 126
Identity and Role-Based Access 126
Real-World Application: Enterprise Application Modernization 127
5.2 Securing Kubernetes Clusters in Azure 128
Securing the Control Plane: Public vs. Private Clusters 130
Hardening the Node Plane and Runtime 130
Enforcing Workload-Level Security Policies 132
Secret Management and Identity Integration 132
Network Segmentation and Zero Trust Access 133
Real-World Application: Secure Healthcare AKS Cluster 133
5.3 Autoscaling and Performance Tuning AKS Clusters 134
Cluster Autoscaler: Scaling Nodes Based on Pod Scheduling Pressure 137
Horizontal Pod Autoscaler (HPA): Scaling Application Instances Based on Metrics 137
Vertical Pod Autoscaler (VPA): Right-Sizing Container Resource Requests 139
Node Pool Strategies and Cost-Aware Scaling 140
Application Performance Tuning Best Practices 140
Case Study: Dynamic Scaling for a Real-Time Sports App 141
5.4 GitOps and CI/CD Pipelines for AKS 141
CI/CD Pipelines: Declarative Delivery Through Azure DevOps and GitHub Actions 142
GitOps: Pull-Based Deployment Using Flux and Argo CD 143
Deployment Strategies: Canary, Blue/Green, and Progressive Delivery 144

TABLE OF CONTENTS

    Managing Secrets and Environment-Specific Configurations 145

    Real-World Application: Platform Engineering at Scale 146

5.5 Azure Container Registry Strategy for Enterprise AKS Deployments 146

    Enforcing Base Image Governance Across Microservices.. 148

    Immutable Tagging and GitOps Workflows 150

    VMSS Runtime Hygiene with crictland Docker gRPC 150

    ACR-Driven Cleanup, Retention, and Policy Enforcement.. 151

    Secure Runtime Operations and Zero-Day Response.. 152

5.6 Designing Hub-and-Spoke Network Topology for Enterprise-Grade AKS Deployments...... 152

    Hub Network: Centralized Control and Security Services... 153

    Spoke Network: AKS Cluster and Private Resource Integration 154

    Traffic Flow and Routing Overview 154

    Advantages of the Hub-and-Spoke AKS Network Design... 155

5.7 Summary.. 156

Chapter 6: Observability, Cost Optimization, and Compliance............................ 157

6.1 Logging and Monitoring with Azure Monitor and Log Analytics... 158

    Azure Monitor: The Control Plane of Observability 161

    Log Analytics Workspaces: Queryable, Correlated Telemetry 162

    Metrics, Alerts, and Visualization.. 162

    Application Insights: Distributed Tracing for Modern Apps 163

    Real-World Application: Centralized Monitoring for Global Enterprise 164

6.2 Cost Optimization Strategies for Azure Workloads... 167

    Azure Cost Management + Billing: Visibility, Analysis, and Forecasting............................ 170

    Azure Advisor: Proactive Cost Recommendations 170

    Reservations and Savings Plans: Committing to Predictable Workloads........................... 171

    Spot VMs and Ephemeral Resources 172

    Hidden Levers for Cost Efficiency.. 172

    Storage and Network Optimization... 174

    Cost Governance with Policies and Tags 175

    Cost Optimization: Multicloud Strategies and FinOps Perspective 176

6.3 Implementing Azure Policies and Security Center 176
    Azure Policy: Defining and Enforcing Governance at Scale 180
    Remediation and Compliance Tracking 181
    Microsoft Defender for Cloud: Posture Management and Threat Detection 182
    Integration with DevOps and Infrastructure As Code 183
    Real-World Application: Achieving PCI-DSS Readiness in Azure 183
6.4 Compliance Considerations for Regulated Industries 184
    Azure Compliance Offerings and Certifications 188
    Key Compliance Design Considerations 188
    Compliance Tooling in Azure 190
    Compliance: Leveraging DINE Policies for Proactive Governance 191
    Real-World Application: Compliance-First Azure Platform for Healthcare 192
6.5 Summary 192

Chapter 7: Real-World Case Studies and Future Trends 195
7.1 Case Study 1: Large-Scale Enterprise Migration to Azure Financial Services at Scale 196
    Business Drivers and Transformation Goals 196
    Designing Scalable Landing Zones 197
    Implementing the Hub-and-Spoke Network Model 201
    Identity and Access Management 201
    CI/CD and Infrastructure as Code 202
    Migration Workloads: From On-Premises to Azure 204
    Security, Risk, and Compliance Controls 207
    Lessons Learned and Outcomes 207
7.2 Case Study 2: Azure for National Health Systems—A Pandemic Response at
    Planetary Scale 208
    The Urgency of Time and the Constraints of Policy 208
    Architecture Overview: A Nation-Sized Microservices Platform 209
    Secure Identity Federation and Pseudonymized Data Design 212
    Integration with Legacy Systems via Hybrid Models 213
    Observability and Resilience Engineering 213
    DevOps at Pandemic Speed 214

TABLE OF CONTENTS

Governance, Audit, and Compliance Assurance 216

Outcomes and Transformational Impact 217

7.3 Case Study 3: Azure for Ecommerce Scale—A Global Retail Platform's Peak Readiness Strategy 217

From Monoliths to Microservices: A Mandate from the CTO 218

Azure Architecture for Scale, Speed, and Safety 218

Real-Time Personalization with Azure AI 222

Observability for Peak Readiness 223

Infrastructure as Code and Deployment Strategy 223

Business Impact 224

7.4 Case Study 4: Financial Services on Azure—Building a Real-Time Risk Engine for Global Trading 224

The Mission-Critical Imperative of Risk at Millisecond Speed 224

Legacy Bottlenecks and Azure Motivation 225

Architecture Overview: A Streaming-Based Risk Engine on Azure 225

Zero Trust, High Compliance Architecture 227

Performance Engineering for Millisecond Analytics 228

Continuous Deployment with Guardrails 229

Business Outcome and Regulatory Confidence 229

7.5 Case Study 5: Healthcare on Azure—Modernizing Electronic Health Records (EHR) with FHIR APIs and Data Interoperability 230

The Healthcare Mandate: From Siloed Systems to Interoperable Care 230

Strategic Vision: FHIR-First, API-Enabled, Cloud-Native 230

Reference Architecture: Azure FHIR Hub with Smart App Integration 231

Ensuring Privacy and Compliance: HIPAA, GDPR, and Beyond 233

DevOps for Digital Health: From Sandbox to National Rollout 234

Outcomes: Healthier Ecosystems, Healthier Patients 234

7.6 Case Study 6: Global Retailer's Multi-region Azure Architecture for High Availability and Edge Acceleration 235

The Digital Shelf Is the Storefront: Why Retail Can't Tolerate Downtime 235

Core Design Pillars: Resilient, Localized, and Observed 236

Architecture Blueprint: Global Load Balancing with Active-Active AKS Clusters 237

Canary and Blue-Green Deployments at Scale 239
Edge Acceleration and Personalization 239
Security and Compliance Across Borders 240
Results: Business Impact and Operational Lessons 240

7.7 Case Study 7: Cloud-Native Banking—High-Frequency Trading (HFT) on Azure
    Confidential Compute 241
Speed, Security, and Secrecy: The Mandate of Modern Trading Platforms 241
Security Beyond Encryption: Shielded Logic in the Cloud 241
Data Flow: From Market Feed to Order Execution in 15ms 242
Observability in a Black Box World 244
Failover Strategy and Circuit Breakers 244
Strategic Outcomes and the Path Ahead 245

7.8 Case Study 8: Sustainability and GreenOps in Azure Enterprise Deployments 246
The Carbon Mandate: From Reporting to Architectural Responsibility 246
Architecture Patterns for Sustainability 246
Instrumentation and Visualization 249
Policy As Guardrail: Carbon-Aware Governance at Scale 250
Organizational Change: Embedding GreenOps into DevOps DNA 250
Outcomes and Metrics 251

7.9 Edge + AI at Scale: Intelligent Applications with Azure Percept and Azure Arc 251
Introduction: Intelligence Where Data Is Born 251
Enterprise Use Case: Predictive Quality Control in Global Smart Factories 252
Architectural Blueprint: Azure Edge AI with Arc and Percept 253
Model Life Cycle and Deployment with Azure ML and IoT Edge 254
Security, Reliability, and Offline Operation 254
Results and Measurable Impact 255

7.10 Lessons Learned: Common Anti-patterns in Azure Transformations 255
Introduction: The Cost of Missteps in Cloud Journeys 255
Anti-pattern 1: Lift-and-Shift Without Re-platforming 256
Anti-pattern 2: Subscription Sprawl Without Management Group Strategy 256
Anti-pattern 3: Overreliance on Portals, Underinvestment in IaC 257
Anti-pattern 4: Ignoring Azure Regions and Availability Zones Strategy 258

TABLE OF CONTENTS

    Anti-pattern 5: Skipping Identity and RBAC Foundations 258

    Anti-pattern 6: Neglecting Observability and Telemetry 259

    Anti-pattern 7: Over-isolation by Region and Network Segmentation 260

    Closing Reflections: The Discipline of Cloud Maturity 261

7.11 Future Trends in Azure Infrastructure and Architecture 261

    Introduction: Evolving Beyond Infrastructure 261

    AI-Native Infrastructure 262

    Confidential Computing and Zero Trust Fabric 262

    Sustainable and Carbon-Aware Architecture 263

    Edge, 5G, and Spatial Computing 263

    GitOps, Platform Engineering, and Internal Developer Platforms (IDPs) 264

    Serverless Infrastructure and Event-Driven Meshes 265

    Conclusion: The Architect As Futurist 265

7.12 Closing Reflections: Architecture Beyond the Horizon 266

7.13 Summary 266

Index 269

About the Author

Roshan Gavandi is a seasoned Azure Solutions Architect and Enterprise Cloud Strategist with over 14 years of hands-on experience designing and delivering large-scale, secure, and resilient cloud infrastructure. His career spans global financial institutions, Fortune 500 enterprises, and high-growth startup organizations that demand uncompromising scalability, regulatory compliance, and operational excellence.

Specializing in Microsoft Azure infrastructure, DevOps automation, and enterprise security, Roshan has led mission-critical cloud transformation programs that modernized legacy systems, implemented Zero Trust architectures, and optimized cloud economics through Infrastructure as Code (IaC) practices using Bicep and Terraform. His expertise spans across the Azure ecosystem, including Kubernetes (AKS), Azure DevOps, hybrid networking, high availability, and multi-region disaster recovery.

Roshan's architecture approach is rooted in automation-first principles and security-by-design thinking, enabling platform engineering teams to build cloud environments that are scalable, observable, and governance-ready. Whether enabling GitOps workflows, enforcing compliance policies through Azure Policy, or deploying enterprise-grade Landing Zones, he is deeply committed to operational excellence and cloud-native reliability.

In addition to his implementation work, Roshan is a passionate thought leader and mentor. He regularly shares his knowledge through public speaking, technical writing, and contributions to the open source community. His blog roshancloudarchitect.me features in-depth articles, best practices, and reference architectures that support DevOps engineers and cloud architects in their daily work.

To follow his work or connect professionally, visit https://www.linkedin.com/in/roshan-gavandi-6b9b79149.

About the Technical Reviewer

Siri Varma Vegiraju is a Senior Software Engineer at Microsoft Azure, where he leads the Network Isolation team securing over 20 global Azure data centers. With deep, hands-on expertise in cloud infrastructure, security, and developer tooling, Siri builds mission-critical systems that protect some of the world's most sensitive commercial and government workloads.

An influential voice in the cloud-native ecosystem, Siri is a core contributor and reviewer for Dapr, a CNCF project used by 40,000 organizations worldwide to build resilient distributed applications. His contributions span both the Java and .NET SDKs, helping shape developer experiences at scale.

A recognized cloud security expert, Siri regularly publishes on DZone, IBM Developer, CNCF, and Cloud Native Now, bringing practical insight to topics like API security, DevSecOps, and agentic infrastructure. He is also a sought-after speaker, delivering technical sessions at global conferences including DeveloperWeek, IEEE Cloud Summit, and Cloud Security Summit.

Introduction

In today's digital economy, infrastructure is no longer a passive foundation; it is the backbone of innovation. Enterprises across the globe are under immense pressure to build systems that are not just operational, but resilient, secure, and scalable from day one. The rise of cloud-native architectures, zero-trust security models, and DevOps-first cultures has dramatically reshaped the expectations from modern infrastructure teams. In this landscape, Microsoft Azure has emerged as a strategic platform, empowering organizations to transition from legacy monoliths to adaptive, modular, and intelligent ecosystems.

This book, *Building Enterprise-Ready Azure Infrastructure: Learn Security, Scalability, and Reliability with IaC & DevOps*, is designed to bridge the gap between Azure documentation and real-world architectural practices. It distills years of field experience into patterns, principles, and production-grade blueprints that can guide architects, DevOps engineers, and cloud strategists alike. Each chapter addresses a core pillar of enterprise infrastructure be it high availability, multi-region DR, secure DevOps pipelines, cost governance, or network architecture and unpacks it with clarity, depth, and implementation-ready examples using tools such as Bicep, Terraform, Azure DevOps, and GitHub Actions.

You will not find mere service overviews or theoretical definitions here. Instead, this book walks you through what it truly means to **design for scale, automate for compliance, and deploy with confidence** in an Azure-first environment. From defining Landing Zones and managing subscription hierarchies to engineering highly available Kubernetes workloads and securing Infrastructure as Code pipelines, this book takes a holistic approach to building cloud infrastructure that aligns with the needs of regulated industries, global platforms, and mission-critical workloads.

Whether you are modernizing a financial institution's infrastructure, setting up a secure supply chain pipeline for healthcare workloads, or scaling a SaaS platform across geographies, this book serves as your blueprint, checklist, and companion in that journey.

Let's build infrastructure that isn't just cloud-hosted but cloud-native, cloud-resilient, and enterprise-ready.

CHAPTER 1

Introduction to Azure Infrastructure

Understanding Cloud Foundations for Enterprise-Grade Environments

It wasn't long ago that the enterprise IT landscape was governed by physical data centers, heavily customized servers, and monolithic systems stitched together by bespoke integrations. Uptime was measured in quarterly reports, not by real-time dashboards. Security was defined by firewalls and badge access to data center cages. In that world, infrastructure was capital procured, racked, and slowly evolved over time. But in today's digital-first economy, where velocity is the new currency and disruption comes from startups as often as from global competitors, infrastructure has transformed from a static asset into a dynamic service layer.

Microsoft Azure, as a leading cloud platform, offers a vast ecosystem of services that power modern enterprises from nimble startups to global banks. But merely lifting and shifting workloads into Azure doesn't guarantee scalability, security, or operational excellence. To harness the full potential of cloud-native infrastructure, organizations must rethink how they design, deploy, and manage their environments. Azure demands a new way of thinking where infrastructure is defined in code, compliance is embedded in pipelines, and reliability is engineered through redundancy and automation.

This chapter sets the foundation for building such enterprise-ready infrastructure on Azure. It explores the strategic shift toward cloud adoption, demystifies the core services Azure provides, and introduces the principles of designing scalable, secure, and compliant cloud architectures. Whether you're architecting for a financial institution governed by strict regulations or enabling rapid growth in a retail SaaS startup, the concepts discussed here will serve as your blueprint.

CHAPTER 1 INTRODUCTION TO AZURE INFRASTRUCTURE

1.1 Cloud Adoption Trends and Enterprise Use Cases

Cloud is no longer just a destination; it has become the operating model of modern business. For most organizations, the question is no longer *if* they should adopt the cloud but *how*. Cloud computing is now integral to digital transformation strategies across industries, enabling innovation at speed, global reach, and elastic scalability. In this context, Microsoft Azure plays a pivotal role as both a cloud platform and an operational model that shapes enterprise IT decisions.

The early wave of cloud adoption was often driven by cost-saving goals, offloading hardware maintenance, reducing capital expenditure, and avoiding over-provisioning. However, enterprise motivations have matured. Today, cloud is embraced for agility, global scalability, regulatory resilience, and the ability to support modern workloads like AI, IoT, and microservices. According to **International Data Corporation** and Gartner, over 90% of large enterprises operate in hybrid or multicloud models, and Azure consistently ranks among the top choices for enterprises seeking both performance and compliance.

Consider the case of a multinational healthcare provider operating across multiple continents. For them, the challenge is twofold: ensure compliance with regional data protection laws such as HIPAA and GDPR while also enabling cross-border access to life-saving analytics and patient services. Azure provides geo-redundant storage, region-specific data residency controls, and built-in compliance blueprints (soon to be deprecated, but replaced with Azure Policy initiatives and landing zones) that make such architectures possible.

In contrast, a fintech startup building a new digital bank may choose Azure for its deep integration with CI/CD tooling, Zero Trust security model, and global peering networks. Here, Azure acts as an innovation enabler, allowing the bank to scale from MVP to millions of users without re-architecting its stack. Their journey may begin with Infrastructure as Code pipelines in Terraform and scale into containerized deployments on AKS with policy enforcement through Azure Policy and Azure Defender for Cloud.

Even government institutions are embracing Azure's hybrid capabilities via **Azure Stack Hyperconverged Infrastructure** and Azure Arc, enabling consistent management across on-premises and cloud environments. These aren't isolated success stories; they're emblematic of a broader shift: infrastructure is now expected to be programmable, policy-driven, resilient, and observable by design.

In these varied use cases, a pattern emerges. Whether in finance, healthcare, manufacturing, or retail, the path to enterprise readiness on Azure depends on three pillars: secure foundations, automated provisioning, and resilient architecture. These pillars are explored throughout this book, beginning with a deep understanding of Azure's core services, which we will now examine in the next section.

1.2 Overview of Azure Core Services (Compute, Storage, Networking)

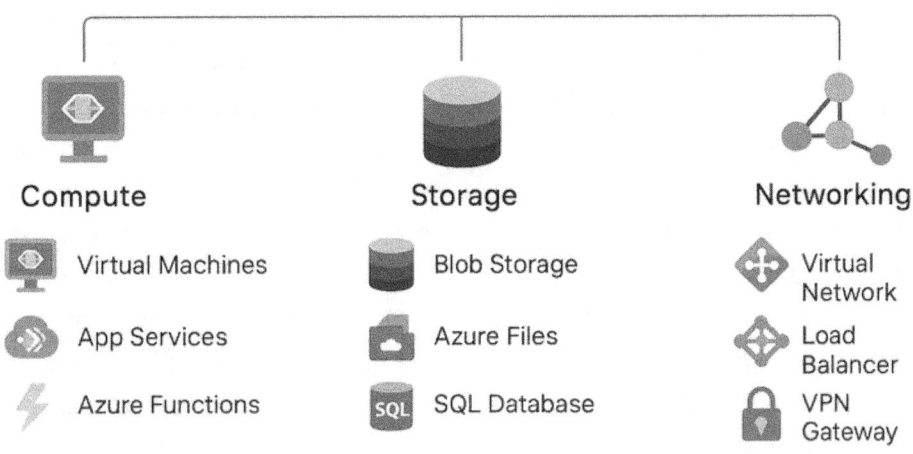

At the heart of every cloud-native architecture lies a trinity of foundational services: compute, storage, and networking. These pillars form the substrate upon which all higher-order Azure capabilities are built, whether it's a serverless application, a machine learning pipeline, or a globally distributed microservices platform. Understanding how these services work and how they can be composed, scaled, and secured is essential to architecting resilient and efficient enterprise environments on Azure.

Compute: The Engine Room of Azure

Compute in Azure is not a monolithic concept; it is an expansive suite of capabilities, each designed for specific use cases and workload profiles. At the most basic level, we have Azure Virtual Machines (VMs), which provide Infrastructure as a Service

(IaaS). These are essentially virtualized servers that can run Windows or Linux and are deployed within a virtual network (VNet). VMs are best suited for legacy workloads, line-of-business applications, or situations where full control over the OS and runtime is needed. They support features like VM Scale Sets (VMSS) for autoscaling and Availability Sets or Zones for high availability.

Yet, Azure's compute model extends far beyond traditional VMs. Azure App Service offers a Platform as a Service (PaaS) option for hosting web applications, RESTful APIs, and background jobs without managing the underlying OS. App Services scale automatically, integrate seamlessly with CI/CD pipelines, and support deployment slots for safe blue/green deployments.

For containerized workloads, Azure provides a progressive compute abstraction model that allows teams to choose the right level of control versus simplicity. At the foundational layer, **Virtual Machines (VMs)** offer full control over the operating system and runtime. **Azure App Service** abstracts the VM layer entirely, allowing developers to deploy web apps and APIs without managing infrastructure. For teams looking to modernize further, **Azure Container Apps** provides a serverless, event-driven container environment ideal for microservices, APIs, and background processing without the operational overhead of managing Kubernetes. It supports features like revision-based deployments, autoscaling (including KEDA-based event triggers), and built-in integration with **Azure Monitor**, **Log Analytics**, and **Azure Defender for Containers** for observability and security.

For advanced orchestration needs, **Azure Kubernetes Service (AKS)** offers the full power and flexibility of Kubernetes while offloading the control plane management to Azure. AKS is best suited for teams that require granular control over container orchestration, custom networking, and stateful workload management. This is where complex microservice architectures, service meshes, and AI inference engines can thrive at scale, but it comes with a steeper operational learning curve. Azure Monitor, as part of the unified observability fabric, is consistently available across all these compute models, whether you're running VMs, App Services, Container Apps, or AKS, ensuring centralized insights regardless of the underlying platform.

For structured and semi-structured data workloads, Azure provides a range of managed database services tailored to different use cases. **Azure SQL Database** supports traditional relational models with full SQL support, while **Azure Cosmos DB** offers globally distributed NoSQL capabilities for applications requiring low latency and high availability at scale. Additionally, **Azure Table Storage** provides a cost-effective NoSQL key/value store suitable for simpler, schema-less data scenarios.

And for event-driven or burst workloads, Azure Functions and Azure Container Apps offer a serverless model where compute resources are automatically provisioned and scaled based on demand, and billing is based solely on consumption. This model is particularly valuable for ephemeral tasks such as data ingestion, alerting workflows, or customer-facing APIs that experience unpredictable traffic patterns.

Storage: The Backbone of Persistence

If compute is the engine, storage is the memory of your architecture. Azure provides multiple types of storage services, each optimized for specific use cases. Azure Blob Storage is the most versatile and widely used ideal for unstructured data such as media files, backups, and data lake scenarios. It supports hot, cool, and archive access tiers, enabling cost optimization based on access frequency. Its immutability features also make it suitable for regulatory compliance in industries like finance and healthcare.

Azure Disk Storage, on the other hand, provides persistent block-level storage for VMs and AKS nodes. Premium SSDs offer high throughput and low latency, while Standard HDDs are cost-effective for less critical workloads. Snapshots and managed disk encryption are built-in, providing options for backup, restoration, and data protection.

For structured and semi-structured data workloads, Azure provides a range of managed database services tailored to different use cases. **Azure SQL Database** supports traditional relational models with full SQL support, while **Azure Cosmos DB** offers globally distributed NoSQL capabilities for applications requiring low latency and high availability at scale. Additionally, **Azure Table Storage** provides a cost-effective NoSQL key/value store suitable for simpler, schema-less data scenarios.

These services are deeply integrated with identity, monitoring, and autoscaling features, allowing enterprise teams to focus on data modeling and query performance rather than infrastructure maintenance.

File-based access is served through Azure Files, which supports SMB and NFS protocols and can be mounted across multiple virtual machines. This is particularly useful for lift-and-shift scenarios, shared configuration storage, or legacy Windows-based applications that require persistent file shares.

Networking: The Fabric That Connects Everything

While compute and storage are tangible, networking is the invisible thread that weaves your infrastructure together. In Azure, all resources, VMs, containers, databases, and functions, reside within one or more virtual networks (VNets). These VNets are logically isolated networks that can span regions, peer with other VNets, and connect to on-premises environments via VPN or ExpressRoute.

Within a VNet, subnets segment workloads in a way similar to departments in an office building. Just as HR and Finance operate in separate rooms with controlled access, application tiers (such as web, API, and database) are segmented into subnets to enforce access control using Network Security Groups (NSGs). NSGs act like virtual firewalls that filter inbound and outbound traffic at both the subnet and NIC level.

When traffic needs to reach the internet or traverse between VNets, Azure provides powerful routing and protection mechanisms. Azure Load Balancer and Application Gateway distribute traffic across back-end pools, while Azure Front Door offers global HTTP/HTTPS routing with integrated CDN and Web Application Firewall (WAF) for modern web applications.

To protect against malicious attacks, Azure DDoS Protection absorbs and mitigates volumetric threats. Meanwhile, Azure Firewall and third-party Network Virtual Appliances (NVAs) provide stateful inspection and traffic logging, critical for regulated industries.

Azure Private Link and Service Endpoints offer secure paths to PaaS services by keeping traffic entirely within the Microsoft backbone, bypassing the public internet. This model is essential for financial institutions and government organizations requiring strict data privacy and isolation.

In sum, Azure's core services are not isolated building blocks; they are deeply integrated, composable components of a cloud-native operating model. Each compute type can connect to the right storage tier and reside within a network architecture that enforces security, redundancy, and compliance. Like the structural steel, plumbing, and electrical systems of a skyscraper, these services form the infrastructure on which every Cloud Native solution is constructed. As we move to the next section, we will explore how to design this infrastructure with intentionality, ensuring it is not only functional but also secure, scalable, and future-ready.

1.3 Designing Scalable and Secure Infrastructure

Designing enterprise-grade infrastructure on Azure is not an exercise in provisioning; it is a continuous, iterative act of architecture. As business needs evolve and technologies advance, this process should be revisited periodically to ensure that your solutions remain robust, secure, and aligned with organizational goals.

The difference is critical. Provisioning is about creating resources; architecture is about shaping the behavior, resilience, and scalability of those resources under real-world conditions. In this section, we shift from foundational services to how they are composed into secure, scalable, and reliable systems.

The Principle of Scalability: Planning for Growth, Not Guesswork

Scalability is not a feature; it is a design imperative. In traditional environments, scale was vertical: when demand increased, you bought a bigger server. In Azure, scale is horizontal, elastic, and predictive. Systems must scale out by adding more instances, not just scale up with more power.

Take a retail enterprise during peak shopping seasons like Diwali or Black Friday. Their website traffic might spike tenfold in a matter of hours. Azure App Service and Azure Kubernetes Service (AKS) allow these workloads to scale automatically based on metrics like CPU utilization, request rate, or custom business indicators such as checkout queue depth. But it's not just about adding instances. True scalability requires

- **Stateless Design**: By stateless, we mean services do not retain information about prior interactions. Each request is self-contained. This enables Azure Load Balancers to distribute traffic evenly and for new instances to join or leave the pool without breaking sessions. State, if needed, is stored in external services like Azure Redis Cache or Cosmos DB.

- **Decoupled Architecture**: By using Azure Service Bus or Event Grid, components are loosely coupled. This ensures that a failure or bottleneck in one part doesn't cascade across the system. It also allows individual components to scale independently.

- **Multi-region Readiness**: For global enterprises, scaling is not just about quantity; it's about geography. Azure Traffic Manager and Azure Front Door enable geo-distribution, routing users to the closest healthy endpoint and minimizing latency. This is especially critical for SaaS platforms with global user bases or financial applications requiring regulatory data residency.

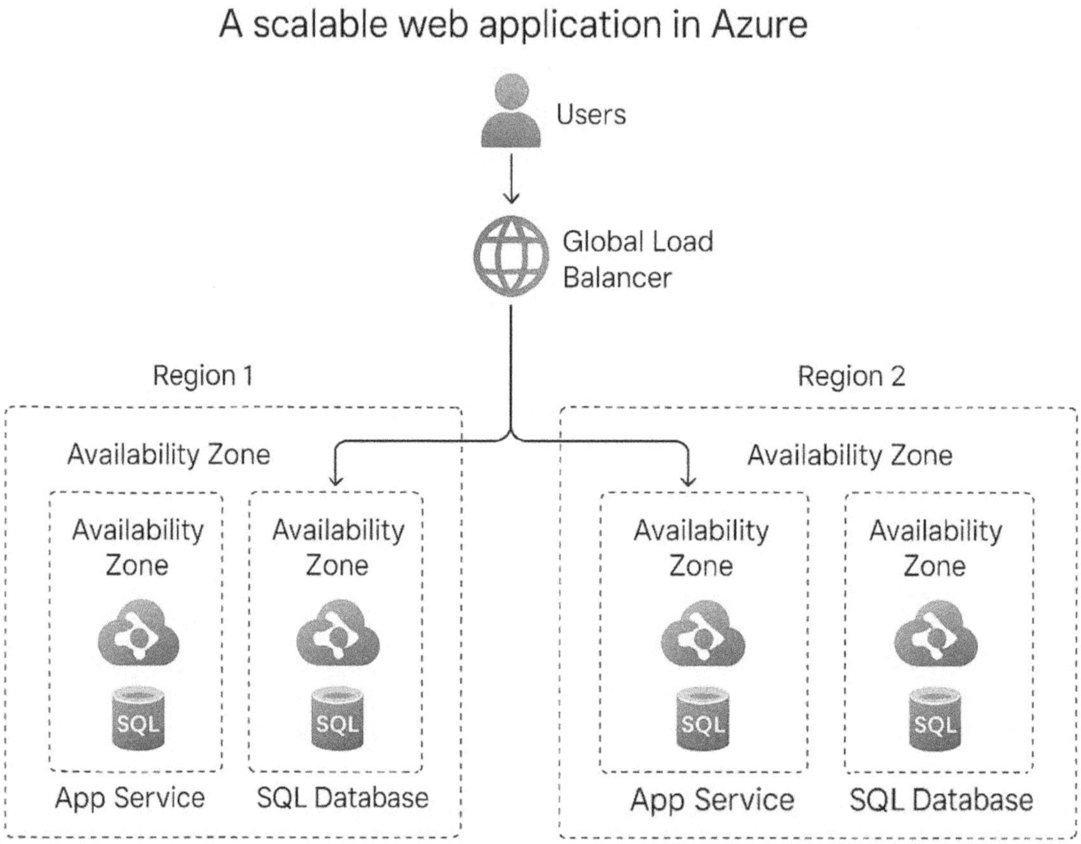

Figure 1-1. Scalable web application architecture in Azure

Figure 1-1 illustrates an example of a scalable web application in Azure, showing how services are distributed across availability zones and regions and fronted by a global load balancer.

The Principle of Security: Built-In, Not Bolted-On

Security must never be an afterthought; it should be woven into every layer of the architecture. Azure follows a shared responsibility model: Microsoft secures the infrastructure, while customers secure their workloads, identities, and access. However, Azure provides rich primitives to implement robust security postures.

Let's begin with **identity**, which is the new perimeter in cloud-native design. Azure Active Directory (Azure AD) underpins access control across Azure services. With Azure AD, enterprises can enforce **Multi-Factor Authentication (MFA)**, **Conditional Access Policies**, and **Privileged Identity Management (PIM)** to ensure that only the right users have the right access at the right time. For service-to-service authentication, **Managed Identities** eliminate the need for secrets in code, allowing services like Azure Functions or VMs to securely access Azure Key Vault, Storage, or databases.

Next is **network isolation**, a foundational principle in secure cloud architecture. Every component in your infrastructure should reside within an **Azure Virtual Network (VNet)**, segmented through well-defined **subnet boundaries** and governed by **Network Security Groups (NSGs)** to control inbound and outbound traffic. To manage this complexity at scale, particularly in multi-subscription or multi-region environments, **Azure Virtual Network Manager (AVNM)** provides a centralized orchestration layer for applying **network topology**, **connectivity configurations**, and **security rules** consistently across VNets. AVNM enables the enforcement of **secure-by-default network policies**, such as automatic blocking of internet egress or noncompliant lateral traffic, making it an essential tool for enterprises adopting Zero Trust and segmented network models in Azure. Services like **Azure Private Link** enable PaaS offerings (e.g., Azure SQL, Azure Storage) to be consumed over private IPs, entirely avoiding the public internet. This is particularly important for organizations in healthcare or government, where strict network isolation is mandated by policy or law.

Encryption is another pillar. Azure provides **encryption-at-rest** by default using platform-managed keys, but enterprises often require **customer-managed keys (CMKs)** or **HSM-backed keys** for regulatory compliance. Similarly, **encryption-in-transit** is enforced via TLS 1.2+ for all Azure APIs and services. Azure Key Vault plays a central role in managing secrets, certificates, and keys integrated directly with compute, storage, and networking layers.

Security also extends to the **application level**. Azure Web Application Firewall (WAF), Defender for Cloud, and Azure DDoS Protection Standard can be enabled at the network edge to detect and mitigate threats before they enter your infrastructure. These services provide protection against common vulnerabilities like SQL injection, cross-site scripting, and denial-of-service attacks while also providing recommendations via Microsoft Defender's Secure Score.

Design Patterns for Enterprise Resilience

Designing for scalability and security often intersects with resilience. Enterprises must architect systems to handle not just load, but failure. This is where **Availability Zones** and **Availability Sets** come into play. For VMs, these options ensure redundancy at the physical datacenter level, protecting against rack failures or hardware outages.

For container workloads, **AKS node pools** can be spread across zones, while persistent storage can be replicated using **Azure NetApp Files** or **Zone-Redundant Storage (ZRS)**. Azure SQL Database and Cosmos DB offer **multi-region replication**, providing both high availability and disaster recovery.

In healthcare, where patient data must be accessible 24×7 and yet protected under stringent laws like HIPAA, these patterns ensure compliance and uptime. In manufacturing, real-time telemetry from IoT devices must continue flowing even during regional outages. Azure's design patterns and reference architectures offer prescriptive guidance for these scenarios.

In summary, designing scalable and secure infrastructure on Azure is about intentional composition, not just of services but of principles. Scalability is achieved through statelessness, elasticity, and geographic distribution. Security is embedded through identity, encryption, network isolation, and continuous monitoring. The art of enterprise architecture is in blending these into solutions that meet both business goals and regulatory demands. As we move into the next section, we explore how Azure's governance and compliance capabilities empower organizations to maintain control as they scale.

1.4 Logical Constructs in Enterprise Azure: Understanding Service Groups

Enterprise adoption of Azure at scale requires more than just technical proficiency; it demands a governance framework that mirrors the complexity of the organization itself. While Azure offers a robust set of built-in hierarchical constructs such as management groups, subscriptions, and resource groups, these are primarily designed to support the cloud's operational and security boundaries. However, enterprises rarely think in those terms. Their mental model of the environment is built around **services**, **business capabilities**, **departments**, and **platforms** that span these technical silos.

This disconnect between **technical hierarchy** and **organizational reality** has given rise to a vital architectural abstraction: the **Azure Service Group (SG)**. A Service Group is not an Azure-native resource in the way a virtual network or storage account is but rather a **logical overlay** that binds related resources together under a common business, functional, or operational identity. It is an enterprise-level pattern that allows architects to represent how systems behave, scale, and are governed, **not just where they live**.

In practical terms, a Service Group might encapsulate a discrete line of business (e.g., SG-ERP for enterprise resource planning), a shared domain (e.g., SG-Identity for authentication and directory services), or even an operational environment (e.g., SG-Production for all customer-facing systems). Crucially, a Service Group can span across multiple **subscriptions**, **regions**, and **resource groups**. This makes it an indispensable construct in large, federated cloud environments where workloads are distributed but must still be managed coherently.

Service Groups are typically implemented using a combination of

- **Tags**: Key/value pairs such as ServiceGroup = Workload-ERP, applied consistently to all resources, enabling cost allocation, policy enforcement, and access tracking.

- **Azure Policy**: Initiatives can be assigned at various scopes, including **management groups**, **subscriptions**, **resource groups**, or individual **resources** to enforce compliance consistently across environments. These initiatives may include policies such as **denying public IP assignments**, **requiring encryption at rest**, or **enforcing tag governance**, ensuring that all security group (SG) related resources adhere to organizational standards.

CHAPTER 1 INTRODUCTION TO AZURE INFRASTRUCTURE

- **Role-Based Access Control (RBAC)**: Access assignments delegated to Azure AD groups named for their SG (e.g., SG-ERP-Owners), aligning operational authority with logical service boundaries.

- **Cost Management scopes**: Budgets, exports, and cost analyses filtered by Service Group tag, making financial governance traceable and actionable.

- **Monitoring and Dashboards**: Observability tools scoped per SG, allowing distinct teams to monitor health, performance, and security posture independently.

The power of Service Groups lies in their ability to **bridge cloud governance with organizational intent**. Instead of managing hundreds of scattered Azure resources, teams can manage a handful of well-defined Service Groups, each governed, monitored, and funded according to its purpose. This abstraction also facilitates standardization; architectural guardrails can be codified per SG, ensuring consistency in logging, encryption, identity integration, and networking patterns across the portfolio.

To illustrate, consider a scenario where an enterprise operates multiple digital services: a B2C ecommerce platform, a back-end ERP system, and an internal HR portal. Each of these workloads may span dozens of resource groups and services, but through the lens of Service Groups SG-eCommerce, SG-ERP, and SG-HR, they can each be tracked, secured, audited, and scaled as a single, logical unit.

CHAPTER 1 INTRODUCTION TO AZURE INFRASTRUCTURE

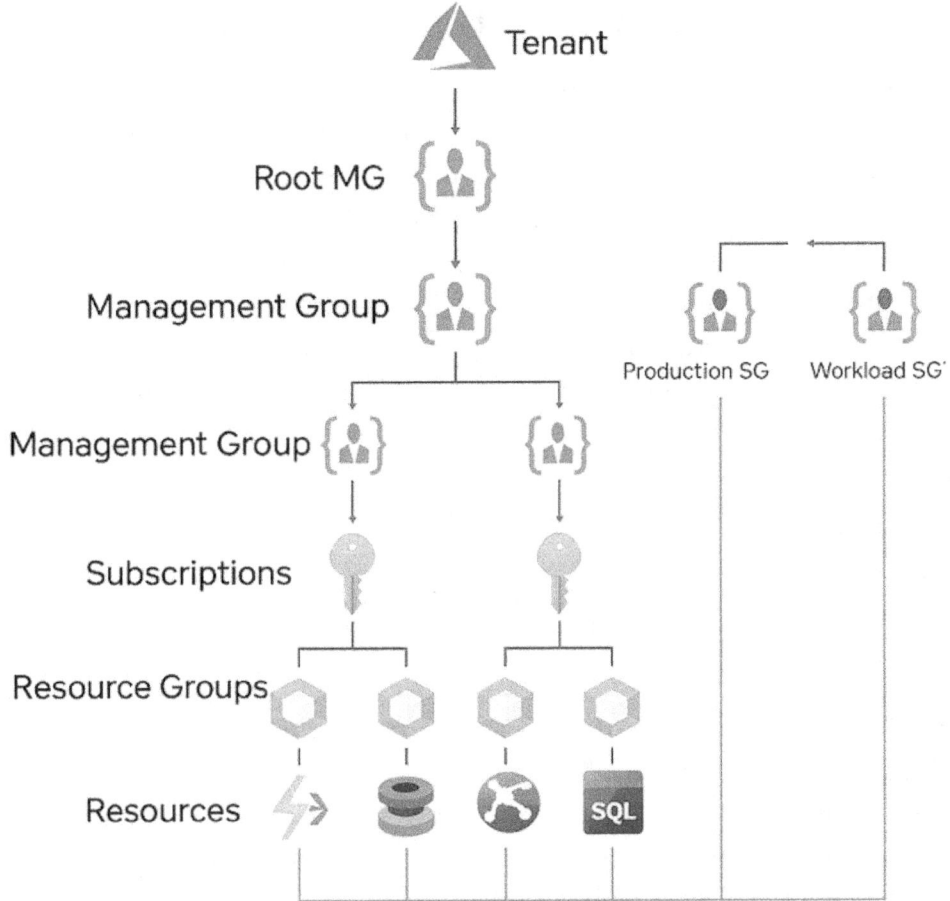

Figure 1-2. *Hierarchical structure for managing cloud resources*

Figure 1-2, placed in the appendix and referenced here, visualizes this abstraction in the context of Azure's resource hierarchy. It demonstrates how SGs map across subscriptions and resource groups and how governance can be enforced at this higher-order boundary.

In summary, **Azure Service Groups are a cornerstone of scalable, business-aligned cloud architecture**. They allow architects to tame complexity by aligning infrastructure governance with service ownership, financial accountability, and operational autonomy. As enterprises grow their Azure footprint, Service Groups offer a unifying pattern that scales with both technology and teams.

1.5 Azure Governance and Compliance

Cloud promises agility, but for large enterprises, unchecked agility can become architectural debt. As organizations scale their Azure footprints across teams, business units, and geographies, governance becomes paramount. Governance is the invisible scaffolding that ensures freedom within a framework: it empowers teams to innovate without compromising security, compliance, or operational consistency.

Governance in Azure isn't a singular service. It is a cohesive framework of capabilities designed to enforce standards, monitor compliance, and maintain organizational integrity at scale. These include tools such as Azure Policy, role-based access control (RBAC), Management Groups, Blueprints (now deprecated and succeeded by Azure landing zone architectures), and Azure Cost Management. Each contributes a layer of control, forming the governance model of an enterprise-grade cloud environment.

Management Groups and Hierarchical Control

Enterprises don't operate as flat hierarchies. Neither should their Azure environments. Azure Management Groups provide a way to structure subscriptions into a logical hierarchy, applying governance policies and access controls at scale. Think of management groups like organizational units in Active Directory; each can represent a business division, geography, or environment tier (e.g., production, staging, dev).

For instance, a global retail company may structure its environment with separate management groups for the Americas, EMEA, and APAC. Under each, subscriptions for workloads like ecommerce, logistics, and analytics are organized. This hierarchy allows central policies to be inherited such as security baselines or allowed Azure regions while enabling local autonomy.

Management Groups are the foundation for scope in Azure Policy and RBAC. Policies assigned at a higher level automatically cascade downward, ensuring consistency and saving operational overhead. This hierarchical governance model is essential for organizations managing hundreds of subscriptions and teams.

Azure Policy: Guardrails, Not Roadblocks

Azure Policy enables the enforcement of compliance standards across resources. It operates by evaluating resource configurations against defined rules called policy definitions and can either deny, audit, or remediate noncompliant resources. Policies are declared in JSON and can be assigned at **multiple hierarchical scopes**, including **individual resources**, resource groups, subscriptions, or management group levels.

For example, in a financial services environment where data sovereignty is critical, a policy can restrict deployment of resources to specific Azure regions like "East US 2" or "UK South." In a healthcare scenario, policies can ensure that all storage accounts have Secure Transfer required or that all virtual machines have Defender for Endpoint enabled.

Azure Policy doesn't just detect noncompliance; it can remediate it. Through "DeployIfNotExists" and "Modify" policy effects, enterprises can auto-enforce tagging, enable diagnostic settings, or deploy agent extensions without manual intervention. This form of **policy-as-code** becomes a critical part of CI/CD pipelines, allowing infrastructure deployments to remain secure and compliant by design.

As of late 2024, Microsoft's shift from Azure Blueprints toward landing zones has further emphasized this model. Landing zones are opinionated, enterprise-ready templates that incorporate policies, RBAC assignments, and network topology from day one. They serve as **pre-baked governance baselines**, ensuring that cloud adoption doesn't sacrifice architectural hygiene.

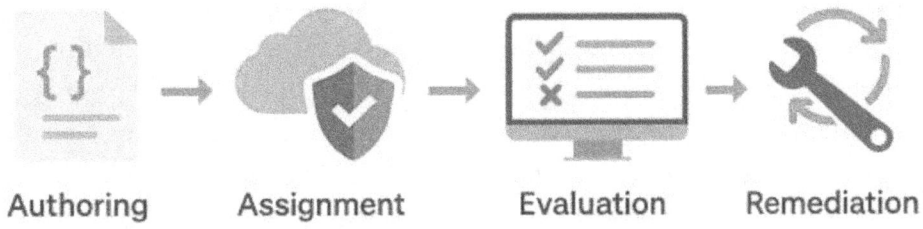

Figure 1-3. *Policy lifecycle flow chart in a cloud environment*

Figure 1-3 illustrates the life cycle of a policy: from authoring and assignment to evaluation and remediation.

Role-Based Access Control (RBAC): Least Privilege at Scale

Access management in Azure is governed by role-based access control (RBAC), which follows the principle of least privilege. By RBAC, we mean users or services are granted only the permissions they need, nothing more. Roles can be built-in (e.g., Reader, Contributor, Owner) or custom-defined with granular actions (e.g., "Can start virtual machines but not delete them").

RBAC works across scopes, resource groups, subscriptions, and management groups and supports both user and service principal managed identities. For large organizations, RBAC should be integrated with Azure AD groups and automated through DevOps pipelines. Assignments should be reviewed regularly, especially for high-privilege roles.

Privileged Identity Management (PIM), part of Azure AD Premium, adds a layer of control for sensitive roles. It allows just-in-time (JIT) role activation, approval workflows, and access expiration critical in sectors like healthcare or government, where audit trails are essential.

RBAC is not just about access; it's about visibility. Overly permissive roles can lead to data exposure or infrastructure drift. Therefore, a robust governance model includes continuous auditing of RBAC assignments, integrated with tools like Microsoft Defender for Cloud and Azure Monitor.

Cost Management and Guardrails for Financial Control

In cloud environments, financial discipline is as important as technical hygiene. Azure Cost Management provides visibility into spending patterns, forecasting, and budgeting. But governance is not just about reporting; it's about enforcing accountability.

Budgets can be scoped per subscription, resource group, or tag and can trigger alerts or automation when thresholds are crossed. Tags like Environment, Cost Center, or Owner are essential for chargeback models in large organizations. Azure Policy can enforce the presence of these tags at deployment time, ensuring financial data is available from day zero.

FinOps financial operations has emerged as a practice to unite engineering, finance, and business teams. In Azure, FinOps is enabled through governance. By embedding cost constraints into the same pipelines that deploy infrastructure, teams operate within guardrails, not after-the-fact reprimands.

Governance for Regulated Industries

For industries like finance, healthcare, and government, governance is inseparable from compliance. Azure offers over 100 compliance certifications globally, including ISO 27001, HIPAA, FedRAMP, and SOC 2. Azure Policy initiatives map directly to these standards, enabling automated audit reporting.

For example, a bank operating in Europe can deploy Azure's "UK Financial Services Authority" initiative, which auto-enforces encryption, monitoring, and network isolation policies aligned with FCA regulations. Similarly, a US-based hospital can apply the "HIPAA HITRUST 9.2" policy initiative across its resource hierarchy.

The key is that governance is codified not in Word documents but in policy definitions, templates, and automation scripts. Compliance is no longer a quarterly report; it is a continuous, machine-verified state of operation.

1.6 Summary

As cloud adoption accelerates across industries, the complexity of managing large-scale Azure environments grows exponentially. In this chapter, we examined why **governance and compliance are not peripheral tasks** to be addressed post-deployment but rather **foundational architectural pillars** that must be embedded from the earliest stages of cloud platform design.

Azure provides a comprehensive governance toolkit **Management Groups, role-based access control (RBAC), Azure Policy, Blueprints, Resource Locks, and Cost Management APIs,** enabling enterprises to define, enforce, and audit the operational and security boundaries of their cloud infrastructure. These tools help organizations align technical implementation with business objectives, regulatory mandates, and internal controls.

One of the core takeaways of this chapter is the **hierarchical and flexible nature of Azure governance**. Policies and role assignments can be applied at the **management group, subscription, resource group,** or even **individual resource level**, enabling fine-grained control and delegated responsibility. The Azure Policy engine, driven by JSON-based definitions, acts as an automated compliance framework that can audit configurations, enforce allowed values, block noncompliant deployments, and apply remediation actions. As a result, compliance is no longer reactive; it becomes continuous and declarative.

We also explored how **governance intersects with cost transparency and accountability**. Through tools like **Azure Cost Management**, **budgets**, and **tags**, organizations can attribute cloud spend accurately, detect anomalies early, and implement guardrails that enforce financial discipline across distributed teams and projects.

In mature environments, governance evolves from being a checklist item to an integral part of the cloud life cycle, **shaping how environments are provisioned, monitored, and secured** while reducing operational risk and technical debt.

As we move forward into Chapter 2, we transition from governance concepts to implementation practices. The next chapter will focus on **Infrastructure as Code (IaC),** the mechanism through which governance, scalability, and security principles are **translated into repeatable, automated deployments**. Using tools like **Bicep**, **Terraform**, and **ARM templates**, we will demonstrate how to codify cloud architecture, enforce compliance by design, and accelerate the delivery of secure, scalable environments in Azure.

CHAPTER 2

Infrastructure as Code (IaC) in Azure

Automating Azure Deployments for Repeatability, Governance, and Velocity

In the traditional IT world, infrastructure was a manual endeavor, servers were racked, cables were labeled, and system administrators used checklists to ensure consistency. While this model worked at a small scale, it became increasingly fragile, expensive, and error-prone as organizations grew. Changes took weeks, documentation lagged behind, and environments drifted from their intended configurations. Infrastructure was not just slow; it was unreliable.

Enter Infrastructure as Code (IaC), a transformational paradigm where infrastructure is declared, versioned, and deployed through code, much like the applications it supports. In Azure, IaC is not merely a convenience; it is a cornerstone of cloud-native architecture. It enables DevOps teams to automate deployments, enforce security and compliance policies, and scale environments consistently across regions, projects, and teams.

This chapter dives deep into the practicalities of implementing IaC on Azure. We will explore how to model infrastructure using Terraform, Bicep, and ARM templates, three powerful and widely adopted IaC tools in the Azure ecosystem. More importantly, we will discuss how to embed IaC into your CI/CD pipelines, integrate it with governance frameworks, and apply best practices for modularity, reusability, and environment consistency in large-scale deployments.

The goal is not just to automate infrastructure creation but to treat infrastructure as a product designed, tested, versioned, and governed through an engineering discipline. Whether you are deploying a multitier application, provisioning Kubernetes clusters, or managing policy-compliant landing zones, IaC is your gateway to reliable and secure enterprise-scale Azure infrastructure.

2.1 Introduction to IaC and Its Benefits

Infrastructure as Code is the practice of defining and managing infrastructure using machine-readable definition files, rather than through physical hardware configuration or interactive configuration tools. But IaC is more than automation; it's about codifying intent, standardizing environments, and treating infrastructure like software.

In Azure, this approach has profound implications. Azure resources, virtual networks, subnets, storage accounts, databases, Kubernetes clusters, and more can all be represented as declarative code. This code can be version-controlled, peer-reviewed, linted, tested, and deployed using the same principles that govern application development.

Why IaC Is Essential in Enterprise Azure Environments

At enterprise scale, manual provisioning is a liability. Teams might use the Azure Portal differently, apply inconsistent naming conventions, forget to enable critical policies, or leave open security holes by accident. IaC eliminates these inconsistencies by establishing a single source of truth for your infrastructure.

Consider a healthcare organization expanding to multiple regions to meet data residency requirements. With IaC, they can define regional variations in configuration like local encryption keys, private endpoints, or monitoring policies while maintaining a core infrastructure baseline. The same Terraform module or Bicep template can be reused with different parameter files, ensuring consistency while allowing customization.

In financial institutions, where regulatory audits demand traceability, IaC enables teams to prove who deployed what, when, and with which configuration. Every deployment becomes an auditable event, stored in Git history or deployment logs. Combined with Azure Policy and Defender for Cloud, IaC makes security compliance a by-product of deployment, not an afterthought.

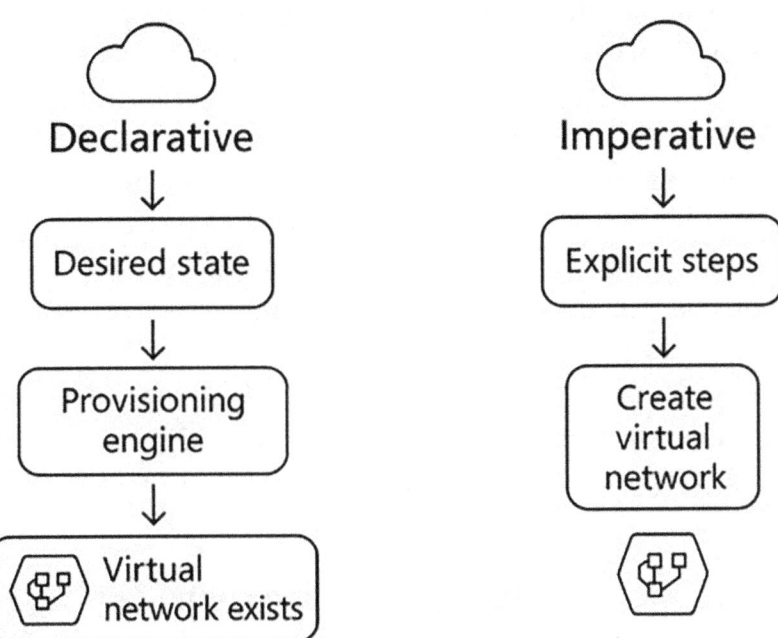

Figure 2-1. *Flowchart comparing imperative and declarative IaC approaches*

Declarative vs. Imperative

IaC on Azure predominantly follows the **declarative model**, where you describe the desired state, and the provisioning engine ensures the actual state matches it. For example, you declare that a virtual network should exist with specific subnets, and the engine makes that true even if it requires creating or modifying underlying resources.

This is distinct from the **imperative model**, where you script a sequence of steps to reach a configuration. Declarative tools like ARM, Bicep, and Terraform are preferred in Azure because they are idempotent; repeated deployments produce the same result, minimizing drift and reducing risk.

Figure 2-1 shows a comparison between imperative and declarative IaC models in Azure and how they impact deployment reliability.

Integrating with DevOps and CI/CD

IaC shines brightest when integrated into DevOps pipelines. Azure DevOps, GitHub Actions, and other CI/CD platforms can trigger infrastructure deployments automatically when a code branch is merged, a release is tagged, or a pull request is approved. This turns infrastructure into a predictable, versioned artifact and ensures changes are tested in lower environments before reaching production.

For instance, a retail SaaS company might have a pipeline where every feature branch spins up an isolated test environment using IaC, runs integration tests, and tears it down after use. This ephemeral environment model reduces costs, accelerates feedback, and increases developer confidence.

By combining IaC with GitOps principles, changes to infrastructure must go through code review and approval workflows just like application code. This enforces **change control**, improves **auditability**, and reduces the blast radius of human error.

2.2 Writing and Deploying Terraform for Azure

Terraform, developed by HashiCorp, is one of the most widely adopted Infrastructure as Code tools in the cloud ecosystem. It's especially popular in organizations managing infrastructure across multiple clouds, thanks to its consistent language, modular architecture, and powerful state management. For Azure environments, Terraform provides a compelling balance between declarative configuration, extensibility, and DevOps integration.

Terraform uses the **HashiCorp Configuration Language (HCL),** a human-readable, domain-specific language designed for clarity and reuse. Its architecture is provider-based, meaning each cloud platform (like Azure) has its own provider plugin that maps HCL syntax to API calls. The Azure provider, azurerm, is mature and actively maintained by both the community and Microsoft, supporting the full spectrum of Azure resources.

The Terraform Workflow: Plan, Apply, and Evolve

Every Terraform execution follows a standard workflow:

CHAPTER 2   INFRASTRUCTURE AS CODE (IAC) IN AZURE

1. **Write**: Define resources using .tf files, typically organized into modules or layers for reusability.

2. **Plan**: Generate an execution plan using terraform plan to preview what will change.

3. **Apply**: Deploy or update resources using terraform apply, aligning real infrastructure with the declared state.

4. **Destroy** (optional): Tear down environments using terraform destroy, useful for ephemeral testing environments.

Let's consider a real example. A financial services company wants to standardize the deployment of a virtual network with three subnets (web, app, and db), a network security group (NSG), and a private endpoint to an Azure SQL Database. Instead of provisioning manually or using separate scripts, they can define all resources in a Terraform module:

```
resource "azurerm_virtual_network" "main_vnet" {
  name                = "vnet-hub"
  address_space       = ["10.0.0.0/16"]
  location            = var.location
  resource_group_name = var.rg_name
}

resource "azurerm_subnet" "web_subnet" {
  name                 = "subnet-web"
  resource_group_name  = var.rg_name
  virtual_network_name = azurerm_virtual_network.main_vnet.name
  address_prefixes     = ["10.0.1.0/24"]
}

resource "azurerm_network_security_group" "web_nsg" {
  name                = "nsg-web"
  location            = var.location
  resource_group_name = var.rg_name

  security_rule {
    name     = "Allow_HTTP"
    priority = 100
```

CHAPTER 2 INFRASTRUCTURE AS CODE (IAC) IN AZURE

```
    direction                  = "Inbound"
    access                     = "Allow"
    protocol                   = "Tcp"
    source_port_range          = "*"
    destination_port_range     = "80"
    source_address_prefix      = "*"
    destination_address_prefix = "*"
  }
}
```

This configuration defines infrastructure declaratively. When committed to a Git repository and tied to a CI/CD pipeline, changes to the code base become traceable, reversible, and reviewable.

State Management and Remote Back Ends

A unique feature of Terraform is its concept of **state**, a snapshot of the infrastructure's last known configuration. This state file allows Terraform to detect drift and reconcile actual infrastructure with the desired configuration. In enterprise settings, storing this state locally is risky. It must be stored in a shared, secure location.

Azure supports **remote state back ends** using Azure Storage accounts. For example:

```
terraform {
  backend "azurerm" {
    resource_group_name  = "tfstate-rg"
    storage_account_name = "tfstateaccount"
    container_name       = "tfstate"
    key                  = "prod.terraform.tfstate"
  }
}
```

Using remote back ends enables **team collaboration**, **locking** (to prevent concurrent updates), and **recovery**, which are all vital in production environments.

In highly regulated industries like healthcare, auditability is essential. Terraform state files should be encrypted, access-controlled, and versioned using blob snapshots. Additionally, organizations often integrate Terraform state life cycle events into audit systems for compliance reporting.

Modules: Building Reusable Infrastructure Constructs

Terraform modules encapsulate related resources and expose input/output variables. For example, a reusable "hub-network" module can encapsulate the entire VNet, subnets, NSGs, and route tables. Teams can consume this module across dev, test, and prod environments with minimal change.

```
module "hub_network" {
  source           = "git::https://github.com/org/modules.git//
                      hub-network"
  vnet_name        = "vnet-hub"
  address_space    = ["10.1.0.0/16"]
  location         = var.location
  environment      = "prod"
}
```

Modules are particularly powerful for building enterprise landing zones and base environments that enforce naming standards, tagging policies, and security controls out of the box. They also support **composition**, allowing infrastructure to scale as applications and teams grow.

Integrating with CI/CD Pipelines

Terraform integrates seamlessly with Azure DevOps and GitHub Actions. A typical workflow includes

- Linting and validating Terraform files with terraform fmt and terraform validate
- Running terraform plan to detect changes on each pull request
- Automatically applying changes to dev/test environments upon merge
- Manual approval gates for production deployments

Here is an excerpt of a GitHub Actions workflow:

```
jobs:
  terraform:
```

```yaml
    runs-on: ubuntu-latest
    steps:
    - uses: actions/checkout@v3
    - name: Setup Terraform
      uses: hashicorp/setup-terraform@v2

    - name: Terraform Init
      run: terraform init

    - name: Terraform Plan
      run: terraform plan -out=tfplan

    - name: Terraform Apply
      if: github.ref == 'refs/heads/main'
      run: terraform apply -auto-approve tfplan
```

This automation ensures that infrastructure deployments are predictable, secure, and repeatable. More importantly, it aligns infrastructure delivery with modern DevOps practices, where infrastructure and application code are deployed as a single, unified pipeline.

In summary, Terraform provides a mature, flexible, and enterprise-ready way to model Azure infrastructure. It empowers teams to standardize, automate, and scale deployments while embedding governance and compliance controls through policy integration, remote state management, and reusable modules. In the next section, we will explore **Bicep**, Microsoft's native IaC language, which offers deep integration with Azure services, a simplified authoring experience, and first-party support for modern deployments.

2.3 Automating Deployments with Bicep and ARM Templates

While Terraform offers cross-platform flexibility and a vast ecosystem, Microsoft's native answer to Infrastructure as Code is found in **Bicep** and **ARM templates,** declarative domain-specific languages built exclusively for Azure. These tools provide first-class integration with the Azure Resource Manager (ARM) API and are ideal for organizations that seek tighter alignment with Azure-native capabilities, faster feature support, and a seamless authoring experience.

The Evolution from ARM to Bicep

Before Bicep, Azure's primary IaC language was the **Azure Resource Manager (ARM) Template**, based on JSON. While powerful and expressive, ARM templates suffered from poor readability, verbosity, and a lack of tooling ergonomics. Simple tasks like looping over subnets or parameterizing resource names required deeply nested expressions that were hard to maintain.

Enter **Bicep**, a language designed by Microsoft to abstract away the complexity of ARM templates while preserving their declarative and idempotent nature. Bicep compiles down to ARM JSON under the hood, but offers a dramatically improved developer experience.

Here's a simple comparison:

ARM Template

```json
{
  "type": "Microsoft.Network/virtualNetworks",
  "apiVersion": "2021-02-01",
  "name": "[parameters('vnetName')]",
  "location": "[parameters('location')]",
  "properties": {
    "addressSpace": {
      "addressPrefixes": [ "[parameters('addressPrefix')]" ]
    }
  }
}
```

Equivalent Bicep:

```
resource vnet 'Microsoft.Network/virtualNetworks@2021-02-01' = {
  name: vnetName
  location: location
  properties: {
    addressSpace: {
      addressPrefixes: [ addressPrefix ]
    }
  }
}
```

The Bicep version is cleaner, shorter, and more readable while retaining full control over resource configuration. Bicep also supports modularity, type checking, loops, conditionals, and tooling extensions in Visual Studio Code, which accelerate adoption among developers and infrastructure engineers.

Bicep Modules and Reuse

Like Terraform, Bicep supports **modularization** through its module keyword. This allows teams to encapsulate repeatable infrastructure patterns such as a network layout, a key vault setup, or an AKS cluster and reuse them across environments.

Consider a Bicep module for a hub VNet:

```
module vnetModule 'modules/hubVnet.bicep' = {
  name: 'deployHubVnet'
  params: {
    vnetName: 'vnet-hub'
    addressPrefix: '10.2.0.0/16'
    location: resourceGroup().location
  }
}
```

Modules are strongly typed and automatically infer parameter types, reducing runtime errors and increasing developer confidence. Additionally, outputs from one module can be passed into another, enabling layered, composable infrastructure deployment.

Deployment Modes and Template Specs

When deploying Bicep or ARM templates, Azure supports two modes:

- **Incremental**: Only creates or updates resources that are not in the desired state. This is the default and safer option.
- **Complete**: Deletes resources not defined in the template. This ensures a strict match between declared and actual state but should be used with caution.

When deploying Azure resources declaratively using Bicep or ARM templates, the Azure Resource Manager (ARM) engine provides two distinct operational modes: **Incremental** and **Complete**. These modes define the behavioral semantics of how the ARM engine evaluates the current state of the resource group in relation to the desired state expressed in the deployment template.

Choosing the correct deployment mode is not merely a syntactic option; it is a **critical architectural decision**. It governs how Azure reconciles drift, manages shared environments, enforces compliance, and balances safety with control. A careless use of the complete mode can result in unintended deletion of critical infrastructure; equally, an overreliance on the incremental mode may lead to configuration drift and hidden infrastructure inconsistencies.

Incremental Mode: Default, Safe, and Evolutionary

Incremental mode is the default behavior in Azure deployments. When a template is deployed using this mode, the ARM engine compares the resources declared in the Bicep or ARM template to the existing resources in the target resource group. It then creates new resources or updates existing ones as necessary. Crucially, any resources that already exist in the resource group but are **not included in the template are left untouched**.

This approach is inherently nondestructive. It aligns with the principles of **idempotent infrastructure updates**, where a deployment can be safely rerun without negative side effects.

Use cases and advantages:

- **Production-grade CI/CD pipelines** where safety and predictability are paramount
- **Shared resource groups**, where multiple teams or systems might provision resources independently
- **Brownfield environments**, where not all infrastructure is fully codified yet and deletions would be risky
- **Progressive modernization**, where templates evolve over time and shouldn't interfere with existing configurations

Operational implications:

- Incremental mode supports iterative updates, making it ideal for Agile and DevOps pipelines.

- However, it does not remove obsolete resources, so **configuration drift can accumulate** over time if not actively managed or audited.

- As such, it's essential to pair incremental deployments with **tagging policies, Azure Policy controls, and resource health monitoring** to ensure ongoing alignment between declared and actual state.

Example CLI usage:

```
az deployment group create \
  --resource-group rg-prod \
  --mode Incremental \
  --template-file main.bicep
```

Complete Mode: Deterministic, Declarative, and Destructive

In contrast, **complete mode** enforces a strict interpretation of the declared template as the **entire authoritative definition of the resource group**. When a template is deployed in complete mode, any resources that currently exist in the resource group but are **not explicitly declared in the template will be deleted** as part of the deployment operation.

This mode ensures perfect convergence between the desired state and the actual state. It embodies the principle of **full declarative ownership**, where the template is treated as the definitive contract for the resource group.

Use cases and advantages:

- **Test or sandbox environments**, where a clean reset is desirable on every deployment cycle

- **Blue/Green infrastructure deployments**, where staging environments must mirror production exactly

- **Highly controlled single-purpose resource groups**, where no external resource contamination is permitted
- **Immutable infrastructure paradigms**, where re-creation is favored over mutation, ensuring zero drift

Operational implications:

- Complete mode introduces **significant risk** in environments where other tools, teams, or processes create resources outside of the template's scope.
- Any untracked or manually created resources will be **irreversibly deleted**, which can lead to outages or data loss if not carefully validated.
- Therefore, complete mode must only be used when the infrastructure boundary is **fully encapsulated and exclusively managed** via IaC.

Example CLI usage:

```
az deployment group create \
  --resource-group rg-test \
  --mode Complete \
  --template-file main.bicep
```

To mitigate the risk, Azure provides a powerful dry-run feature called what-if, which previews the changes without applying them. This is especially important before executing a complete mode deployment:

```
az deployment group what-if \
  --resource-group rg-test \
  --mode Complete \
  --template-file main.bicep
```

Choosing the Right Mode: Strategic Guidance for Architects

The choice between **incremental** and **complete** mode must be deliberate and context-driven. Below is a synthesized strategy for selecting the appropriate deployment mode based on environmental and organizational constraints:

Scenario	Recommended Mode	Rationale
Shared production resource groups	Incremental	Avoids accidental deletion of unrelated resources. Safer for cross-team usage.
Isolated test/staging environments	Complete	Enforces a clean state and removes stale resources. Useful in daily rebuilds.
Immutable infrastructure rollouts	Complete	Guarantees alignment between declared and actual state. Minimizes drift.
Regulated environments with audit controls	Incremental (with policy) or Complete (with approval gates)	Depends on risk tolerance. Combine with Azure Policy and what-if analysis.
Multi-tenant applications using scoped resource groups	Incremental	Required to protect tenant-specific deployments from broad deletions.

For enterprises managing multiple environments or offering reusable infrastructure blueprints to internal teams, **Template Specs** serve as a way to version and distribute Bicep/ARM templates across subscriptions. A template spec is a managed resource that stores your IaC logic and exposes it as a versioned artifact.

Imagine a central platform team creating a standardized AKS deployment spec. Application teams can then deploy compliant clusters using:

```
az deployment group create \
  --template-spec "/subscriptions/xxx/resourceGroups/infra-rg/providers/Microsoft.Resources/templateSpecs/aks-cluster/versions/1.0.0" \
  --parameters clusterName="team-aks" location="eastus"
```

This ensures governance, consistency, and reuse at enterprise scale.

Parameterization and Secret Management

One of Bicep's strengths, as with its underlying ARM templates, is the ability to securely integrate with **Azure Key Vault** for managing sensitive values such as passwords, connection strings, and access keys. While this capability is not unique to Bicep, Bicep simplifies the syntax and parameter declarations for referencing Key Vault secrets. ARM templates also support secret parameters via the reference function and secure object types, enabling the same integration pattern through more verbose JSON structures.

Secrets can be passed into parameters using @secure() annotations and resolved at runtime. This keeps sensitive configurations like database passwords or API keys out of code repositories and CI/CD logs.

```
param adminPassword string {
  metadata: {
    description: 'Admin password for VM'
  }
  @secure()
}
```

This integrates tightly with **Azure DevOps** and **GitHub Actions**, where pipeline variables can pass secrets into Bicep deployments while maintaining auditability.

CI/CD Integration and Linting

Bicep is deeply integrated with Azure CLI and Azure PowerShell. For example:

```
az bicep build --file main.bicep    # Compiles to ARM JSON
az deployment sub create --location eastus --template-file main.bicep
--parameters ...
```

In pipelines, Bicep files can be linted with bicep lint, tested for syntax, and compiled ahead of time. Organizations often define a quality gate in their CI pipelines to fail builds if bicep lint returns warnings or if parameter files fail validation.

A GitHub Actions sample step might look like:

```
- name: Build and Validate Bicep
  run: |
    az bicep build --file main.bicep
    az deployment group validate --resource-group prod-rg --template-file
    main.bicep --parameters @prod.parameters.json
```

These practices embed code quality and runtime assurance into the delivery life cycle key tenets of DevSecOps in Azure.

In summary, Bicep offers a clean, Azure-native, and developer-friendly approach to IaC. It reduces boilerplate, accelerates onboarding, and provides a robust foundation for secure, policy-driven deployments. For teams fully invested in Azure and seeking rapid access to the latest features, Bicep is often the preferred choice. In contrast, ARM templates, while still supported, are more suitable for tooling scenarios or backward compatibility with existing deployments.

As we turn to the final section of this chapter, we will discuss **best practices for implementing IaC in large-scale enterprise environments**, including design patterns, organizational standards, and pipeline strategies that ensure consistency, reusability, and security at scale.

2.4 Best Practices for IaC in Large-Scale Environments

At a small scale, Infrastructure as Code feels like an automation tool script that saves time. But at enterprise scale, it becomes something else entirely: a platform for governance, a foundation for compliance, and an engine for agility. As environments grow to span hundreds of applications, dozens of teams, and multiple regions, IaC must evolve from a set of templates into a structured system that enables scale without sacrificing control.

In this section, we move beyond syntax and tool-specific features and focus on patterns and best practices that make Infrastructure as Code sustainable, secure, and resilient in large-scale Azure environments.

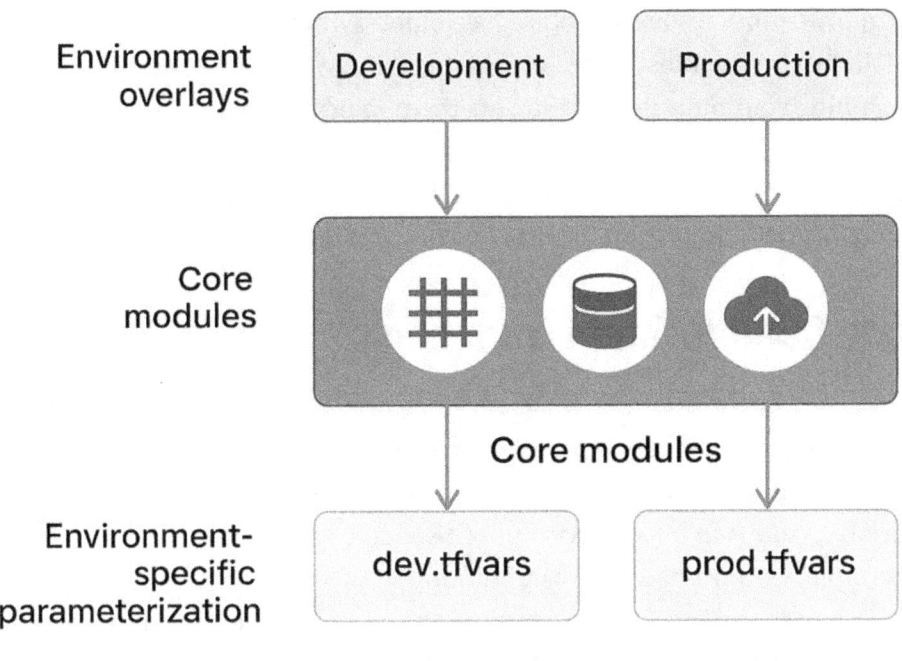

Figure 2-2. Composable IaC architecture with shared core modules and environment overlays

Adopt Modular Architecture: Build Infrastructure Like Software

The first rule of scalable IaC is modularity. Just as software developers break monolithic code bases into modules, packages, or services, infrastructure code should be composed of **reusable, loosely coupled modules**. Each module should encapsulate a distinct concern: networking, identity, monitoring, databases, and expose inputs and outputs in a well-defined contract.

For instance, a "hub-network" module can abstract away all the complexity of a virtual network, subnets, NSGs, and route tables. Application teams can consume this module with minimal knowledge of its internals, passing only what's necessary like CIDR ranges and region names.

In Terraform, modules are structured as folders with main.tf, variables.tf, and outputs.tf. In Bicep, modules are bicep files imported via the module keyword. In both cases, versioning your modules and storing them in private registries or Git repositories allows you to evolve infrastructure safely and predictably.

Figure 2-2 illustrates a composable IaC architecture with shared core modules, environment overlays, and environment-specific parameterization.

Separate Configuration from Logic

Hard-coding values like resource names, secrets, or region codes undermines reusability and security. Adopt the principle of **separation of concerns** by keeping configuration (parameters, variables, secrets) separate from deployment logic. Use parameter files or CI/CD variable groups to inject environment-specific values.

For example, your infrastructure definition for AKS remains the same across dev, qa, and prod environments. Only the number of nodes, SKU, or location might change, supplied through external parameter files or pipeline variables.

This practice also facilitates **environment promotion,** where the same base infrastructure code is applied to each stage of a pipeline with increasing rigor (e.g., validation, testing, approval gates).

Embed Security and Governance from Day Zero

Infrastructure code should not merely provision resources; it should **enforce compliance**. That means including policy assignments, diagnostic settings, logging configurations, and secure access patterns in your templates or modules. Waiting until after deployment to secure infrastructure is a recipe for drift and risk.

For example:

- Every storage account provisioned by your IaC module should have logging enabled, HTTPS-only enforced, and a private endpoint attached.

- Every virtual machine should deploy with Defender for Endpoint enabled and linked to Log Analytics.

- Every key vault should be integrated with RBAC, not access policies, and deployed into a private subnet.

These controls, when baked into reusable modules, eliminate the risk of human omission and help pass security audits with ease.

In Azure, use Azure Policy and Blueprints (or landing zones) in conjunction with IaC to enforce tagging standards, region restrictions, and SKU limits. This ensures that every deployment aligns with corporate governance, regardless of who initiates it.

Use CI/CD Pipelines for Delivery and Drift Detection

Manual deployment of IaC defeats its purpose. Infrastructure code should always be deployed through **continuous integration and delivery (CI/CD)** pipelines, not local terminals. Pipelines provide version control, traceability, automated validation, and integration with change approval workflows.

A robust pipeline will include

- terraform validate or bicep build and bicep lint
- Unit and integration tests using tools like Terratest or Pester
- terraform plan or az deployment validate with change previews
- Manual gates for production approval
- Post-deployment tests or health checks
- Slack/Teams notifications and alerts

In addition to deployment, IaC pipelines should also be used to detect **drift** when actual infrastructure no longer matches the declared state. Tools like terraform plan or az resource list comparison scripts can surface such discrepancies and initiate remediation workflows.

Manage State Securely and Centrally

For tools like Terraform, managing state is not optional; it is foundational. Use remote back ends with Azure Blob Storage and enable state locking via Azure Cosmos DB or built-in blob leases. Encrypt state-at-rest using customer-managed keys (CMK) if required by compliance.

Avoid storing secrets in state files or parameter files. Use Key Vault references or secure pipeline variables instead. Implement RBAC policies on storage accounts and containers that hold state to prevent tampering.

Adopt a GitOps and Change Management Model

GitOps is a natural extension of IaC. All changes to infrastructure must go through Git. Developers submit pull requests, peer reviews are conducted, and CI/CD pipelines apply the changes. This model enforces change discipline and creates a rich, auditable history of how your environment has evolved.

Use branches to manage environments (main, dev, feature/*) and tag releases to trigger specific environment deployments. Integrate tools like Azure DevOps Environments or GitHub Environments to enforce approvals, rollback policies, and identity-scoped secrets per stage.

In regulated environments, where change control boards (CCBs) are mandatory, GitOps provides verifiable artifacts who made the change, what was changed, when it was applied, and whether it passed automated controls.

In conclusion, Infrastructure as Code is not just a deployment technique; it is an architectural philosophy. When implemented with modularity, security, governance, and automation, IaC enables enterprises to scale safely, innovate faster, and operate with confidence. It aligns infrastructure with engineering discipline and transforms infrastructure into a managed, testable, and composable asset.

With this foundation in place, we now shift our focus to a critical aspect of cloud architecture: **networking and security**. In the next chapter, we will explore how to build secure, scalable networks in Azure, applying zero trust principles, hybrid connectivity, and defense-in-depth strategies to protect your cloud infrastructure end-to-end.

2.5 Engineering Multi-region Deployments with Safe Deployment Practices and GitOps in Azure DevOps

The challenge of deploying infrastructure is no longer confined to ensuring a single region is properly configured. In today's global enterprises, infrastructure architects must design systems that are resilient across **multiple Azure regions**, adhere to **regional regulatory standards**, and can **safely evolve over time without disrupting production workloads**. From banking systems that must remain online during regional outages to healthcare applications constrained by data sovereignty laws, the modern infrastructure landscape demands more than automation; it demands intelligent orchestration.

This section addresses that challenge head-on. It explores how cloud engineering teams can leverage Azure DevOps and Bicep to implement **multi-region Infrastructure as Code (IaC)** pipelines that are not only repeatable and scalable but also **safe**, **auditable**, and **declarative**. It combines three critical paradigms:

1. **Matrix Strategy for Multi-region Deployments**: Enabling code reuse and uniformity across global deployments

2. **Safe Deployment Practices (SDP)**: Ensuring changes are rolled out progressively to minimize risk

3. **GitOps-Based CI/CD Workflows**: Anchoring the entire infrastructure life cycle in Git for traceability and compliance

This unified model ensures that infrastructure is not just deployed, but deployed **well** with care, with control, and with confidence.

Why Multi-region IaC Is No Longer Optional

Enterprises that operate at scale inevitably face the requirement to support multiple Azure regions. This may stem from various operational, compliance, or strategic needs:

- **Disaster Recovery (DR)**: Business continuity planning requires workloads to failover from a primary region to a secondary or tertiary location.

- **Latency Optimization**: End users in Asia-Pacific, Europe, and North America each expect low-latency access to applications, necessitating regionally local deployments.

- **Data Sovereignty**: Laws like the European GDPR or India's DPDP Act restrict data storage to within regional borders.

- **High Availability**: Applications must remain available even during zone or regional outages, necessitating redundant infrastructure in other locations.

Historically, teams handled such deployments manually or through duplicated code bases, which quickly grew unmanageable and brittle. Today, Azure DevOps' **strategy.matrix** capability allows infrastructure engineers to deploy across regions using a centralized template and parameterized logic, offering **standardization without duplication**.

Matrix Strategy: Declarative Global Infrastructure with Local Customization

The matrix strategy in Azure DevOps empowers a single job to run multiple times, each with a different set of variables. When applied to infrastructure provisioning, this pattern elegantly solves the challenge of deploying the same Bicep templates across different Azure regions and environments.

Consider the following example:

```
strategy:
  matrix:
    eastus:
      environmentName: 'eastus'
      resourceGroup: 'rg-east-prod'
      bicepFile: 'infra/main.bicep'
      location: 'eastus'
    westeurope:
      environmentName: 'westeurope'
      resourceGroup: 'rg-west-prod'
      bicepFile: 'infra/main.bicep'
      location: 'westeurope'
    centralindia:
      environmentName: 'centralindia'
      resourceGroup: 'rg-india-prod'
      bicepFile: 'infra/main.bicep'
      location: 'centralindia'
```

Each matrix entry allows the job to run with different inputs for location, resourceGroup, and other parameters, but the job logic remains unchanged. This ensures that regional deployments are **consistent in structure**, yet **contextually aware** of local differences.

Introducing Safe Deployment Practices (SDP) in Infrastructure Pipelines

While matrix-based parallelism offers speed and scale, it introduces operational risk. A misconfigured policy, a failed parameter, or an overlooked dependency when deployed simultaneously across all regions can cause a global outage in seconds.

To mitigate this, Microsoft and other hyperscalers advocate **Safe Deployment Practices (SDP)**: a disciplined rollout strategy that limits change exposure and controls the blast radius of failures. In infrastructure pipelines, this translates into **region-by-region deployment sequencing**, often with **approval gates**, **monitoring checkpoints**, and **automated rollback logic**.

Azure DevOps enables this with **multistage pipelines**, where each stage represents a specific region and executes only if the previous region's deployment has succeeded. Here's a conceptual layout:

```
stages:
- stage: DeployEastUS
  displayName: 'Primary Region'
  jobs:
    - job: deployEast
      ...
- stage: DeployWestEurope
  displayName: 'Secondary Region'
  dependsOn: DeployEastUS
  condition: succeeded()
  jobs:
    - job: deployWest
      ...
- stage: DeployIndia
  displayName: 'Disaster Recovery Region'
  dependsOn: DeployWestEurope
  condition: succeeded()
  jobs:
    - job: deployIndia
      ...
```

Each regional deployment becomes a **checkpointed phase**, with optional manual or policy-based approvals:

```
environments:
- name: eastus
  approval:
    reviewers:
      - group: 'CloudOps'
- name: westeurope
  approval:
    reviewers:
      - group: 'ComplianceTeam'
```

This approach transforms the deployment from a fire-and-forget operation into a **controlled and observable release pipeline,** one that can catch and contain failures before they cascade.

GitOps Flow: Commit-Driven Infrastructure Life Cycle

The third pillar of this deployment model is **GitOps**, a practice that treats Git as the single source of truth for infrastructure. In this model, infrastructure is changed by submitting pull requests to a Git repository. Every change is reviewed, tested, built, and then deployed through an automated pipeline, **not manually applied through portals or ad hoc scripts**.

Here's how the flow typically works in an Azure DevOps environment:

1. A cloud engineer makes a change to infra/main.bicep or parameters/prod.bicepparam.

2. A **build pipeline** is triggered upon commit, which

    - Validates the Bicep syntax

    - Optionally performs a what-if preview

    - Publishes the files as pipeline artifacts

3. A **deployment pipeline** is triggered by the successful build, which

    - Downloads the artifact

    - Iterates through a matrix or staged rollout across regions

CHAPTER 2   INFRASTRUCTURE AS CODE (IAC) IN AZURE

- Applies the Bicep templates using az deployment group create.

Sample Build Pipeline

```
trigger:
  branches:
    include: [main]
  paths:
    include: [infra/**]
pool:
  vmImage: 'ubuntu-latest'
steps:
- task: AzureCLI@2
  inputs:
    azureSubscription: 'MyServiceConnection'
    scriptType: 'bash'
    scriptLocation: 'inlineScript'
    inlineScript: |
      az bicep build --file infra/main.bicep
- task: PublishPipelineArtifact@1
  inputs:
    targetPath: 'infra'
    artifact: 'iac-artifact'
```

Sample Deployment Job with Matrix

```
- job: DeployToRegion
  strategy:
    matrix:
      ...
  steps:
  - task: AzureCLI@2
    inputs:
      azureSubscription: 'MyServiceConnection'
      scriptType: 'bash'
      scriptLocation: 'inlineScript'
```

```
  inlineScript: |
    az deployment group create \
      --name deploy-$(environmentName) \
      --resource-group $(resourceGroup) \
      --template-file $(bicepFile) \
      --parameters location=$(location)
```

This separation of concerns **builds what changes, deploys where it matters**, and brings security, governance, and clarity to every infrastructure life cycle.

Real-World Use Case: Safe Rollout for Global FinTech Infrastructure

Consider the example of a FinTech company operating in the United States, the EU, and APAC markets. Their platform runs critical payment services, and any infrastructure error can disrupt millions of transactions.

To ensure safety and compliance, they adopt the following model:

- Deploys first to eastus, their primary production region
- Monitors key metrics via Application Insights and Log Analytics
- Requires an approval from the Site Reliability Engineering (SRE) team before proceeding to westeurope
- Finally, deploys to centralindia, their designated disaster recovery region

Multi-region deployments are a hallmark of cloud maturity, but without **discipline**, they can become liabilities. Through a combination of **matrix-driven automation**, **Safe Deployment Practices**, and **GitOps-based pipelines**, Azure DevOps and Bicep enable enterprises to build and deliver global infrastructure that is not only scalable but also **controlled, compliant, and resilient**.

These patterns form the foundation of enterprise-grade infrastructure delivery. They don't just automate; they **institutionalize safety**, **governance**, and **agility**.

2.6 Summary

Infrastructure as Code (IaC) has emerged as a foundational discipline in the modern cloud operating model. In Azure, the transition from imperative, manual provisioning to declarative, codified infrastructure not only increases velocity and repeatability but also embeds security, compliance, and architectural governance directly into the development life cycle.

This chapter introduced readers to the principles and practices of IaC in the Azure ecosystem, focusing specifically on **Bicep**, the domain-specific language that abstracts away the complexity of JSON-based ARM templates while retaining their full expressive power. Through Bicep, infrastructure declarations become concise, modular, and maintainable, supporting the same robust deployment semantics and secure integrations, such as with Azure Key Vault, but in a more accessible and auditable format.

We then explored how Bicep integrates seamlessly with **Azure DevOps pipelines**, enabling the automation of environment provisioning and updates via CI/CD workflows. Leveraging **pipeline triggers**, **artifact stages**, and the **AzureCLI@2** task, we constructed build and deployment pipelines that are Git-driven and environment-aware.

A key innovation discussed in this chapter is the use of the **matrix strategy in Azure DevOps**, which allows for scalable, region-specific infrastructure rollouts without duplicating code. This strategy not only reduces pipeline complexity but also supports enterprise-wide patterns such as **multi-region availability**, **data residency compliance**, and **disaster recovery preparedness**.

To mitigate the operational risks of large-scale deployments, we introduced the concept of **Safe Deployment Practices (SDP)**. SDP encourages controlled, phased rollouts across regions using pipeline stages, environment checks, and manual or automated approval gates. This method minimizes blast radius and allows issues to be detected and remediated in early phases before full rollout. The combination of matrix logic with SDP principles forms a powerful pattern for global infrastructure resilience.

Finally, the chapter introduced the **GitOps model** for IaC delivery. By treating Git as the single source of truth, changes to infrastructure are version-controlled, reviewed via pull requests, validated in build pipelines, and applied to Azure environments through structured release pipelines. This model enforces traceability, compliance, and reproducibility at scale.

We also delved into the semantics of **deployment modes in Bicep and ARM**, namely, **incremental** and **complete** modes, and provided nuanced guidance on when each should be used. Incremental mode offers a safe, additive approach ideal for production systems, while complete mode enforces strict convergence and is better suited to isolated or disposable environments.

Together, the techniques explored in this chapter equip cloud architects and engineers with the practices and tooling required to manage infrastructure **declaratively**, **securely**, and **at scale**. Whether deploying to a single region or orchestrating a globally distributed landing zone, the combination of Bicep, Azure DevOps, and GitOps workflows forms the backbone of a resilient and modern Azure platform.

In the chapters ahead, we will extend these patterns into the realms of **networking**, **security**, **high availability**, and **container orchestration**, building upon the IaC foundation laid here to deliver full-spectrum production-ready Azure infrastructure.

CHAPTER 3

Azure Networking and Security

Designing Secure, Scalable, and Resilient Network Architectures in the Cloud

Networking and security are often treated as adjacent concerns, but in a cloud-native architecture, they are inseparable. In Azure, the boundaries of your network define your security posture. Every virtual machine, container, storage account, and application endpoint you deploy lives within a virtual network, an abstraction that replaces the traditional firewall and switchboards of on-premises data centers. But while Azure's networking model simplifies many physical constraints, it introduces new layers of abstraction and complexity. Subnets are no longer just IP ranges; they are security domains. Firewalls are programmable. Routing tables are dynamic. And with the rise of hybrid and multicloud deployments, network perimeters have become porous and context-driven.

Security, too, must evolve. Traditional perimeter-based defenses are no longer sufficient. In a world where users work from anywhere, data resides everywhere, and applications scale dynamically, the only viable security model is **Zero Trust**: never trust, always verify. Azure supports this through a rich set of services, identity-driven access controls, network isolation, encryption at every layer, and intelligent threat detection.

This chapter will guide you through the architecture of secure, scalable networking in Azure. You will learn how to structure virtual networks, design hybrid connectivity, enforce access controls with NSGs and Azure Firewall, and implement Zero Trust principles with Private Endpoints and identity-aware access. Real-world examples from fintech to healthcare will illustrate how these principles are applied in regulated, high-performance environments. By the end of this chapter, you will understand how to design Azure network topologies that are not only efficient but also resilient to attack, failure, and change.

CHAPTER 3   AZURE NETWORKING AND SECURITY

3.1 Virtual Networks, Subnets, and Private Endpoints

Azure Virtual Network (VNet) is the fundamental building block of Azure networking. It represents a **logical isolation boundary** within the Azure fabric, akin to a private data center within the public cloud. Every compute resource whether it's a virtual machine, container instance, or a private PaaS service must live within a virtual network if it is to be securely addressable, routable, and governable. As such, the way you design VNets and subnets has a profound impact on your overall infrastructure security, performance, and maintainability.

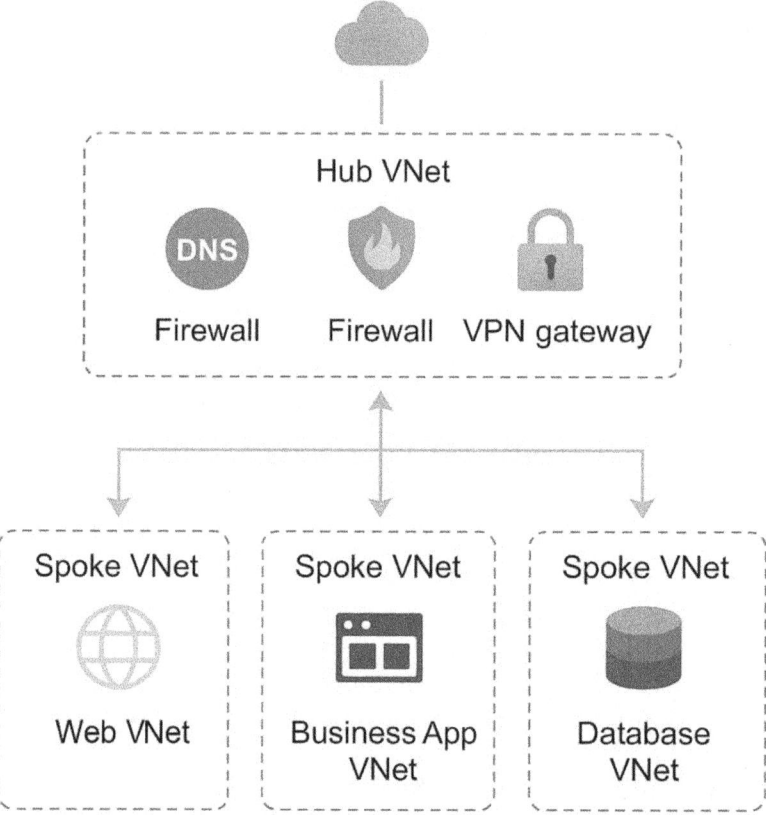

Figure 3-1. Hub-and-spoke network architecture with a centralized Hub VNet (DNS, firewall, VPN gateway) connected to Web, Business App, and Database Spoke VNets

Designing Azure Virtual Networks: Your Logical Data Center

A VNet is defined by an address space using CIDR (Classless Inter-Domain Routing) notation, such as 10.0.0.0/16, and may span multiple availability zones within a region. Within this space, you create **subnets**, logical partitions that segment workloads, enforce access control, and facilitate routing.

Think of a VNet as an office building and subnets as departments on different floors. The HR department (subnet) should not have unrestricted access to the Finance department (another subnet). Similarly, your front-end web servers might reside in a subnet-web, your application logic in subnet-app, and your databases in a locked-down subnet-db, all within a single VNet.

VNets can **peer** with each other within the same region or across regions using **VNet Peering**. Peered VNets communicate via the Azure backbone, not over the public internet, resulting in low latency and high throughput. Peering is non-transitive: if VNet A peers with B and B peers with C, A cannot communicate with C unless a direct peer is established.

In large enterprise deployments, VNets are often structured into **hub-and-spoke topologies** (Figure 3-1). The hub VNet contains shared services, DNS, firewalls, and VPN gateways, while spokes are application-specific VNets peered to the hub. This architecture simplifies security enforcement, centralizes policy control, and enables scalable segmentation.

Subnets and Network Segmentation

Subnets in Azure are more than just IP boundaries; they are **security zones**. Each subnet can be associated with **Network Security Groups (NSGs)** that act like stateless firewalls, controlling inbound and outbound traffic based on rules defined by source/destination IPs, ports, and protocols. By isolating workloads into different subnets, you apply the principle of least privilege at the network layer.

For example, your web subnet may allow inbound HTTP/HTTPS traffic from the internet but restrict all outbound traffic except to the application subnet. The app subnet can then communicate with the database subnet but block all other traffic. Subnets can also be associated with **Route Tables**, allowing you to override default routing and direct traffic to firewalls or virtual appliances for inspection.

It's worth noting that some Azure services like Azure Kubernetes Service (AKS) require specific subnet configurations for pod and node pools. Planning IP ranges carefully in early design stages avoids future IP exhaustion or disruptive changes during scaling.

In multi-tenant SaaS environments, it's common to assign each tenant a dedicated subnet or even a VNet, further improving isolation and compliance.

Private Endpoints: Securely Consuming PaaS Services

A major shift in cloud networking is the move from public endpoint exposure to **private consumption of services**. Azure Private Link enables this by assigning **Private Endpoints** a private IP address from your VNet to supported PaaS services like Azure Storage, Azure SQL, Cosmos DB, Key Vault, and many others.

Private Endpoints allow traffic to flow **entirely within the Azure backbone**, never traversing the public internet. This reduces exposure to attacks, simplifies firewall rules, and helps meet regulatory requirements in sectors like healthcare and finance.

Consider a healthcare application storing patient records in Azure SQL. Without Private Endpoints, the SQL server would have a public IP, and you'd need to allowlist client IPs, risking data leakage through misconfiguration. With Private Endpoints, SQL is reachable only through a private IP from within the app's subnet, eliminating the need for any public exposure.

Private Endpoints support **DNS integration** with private zones, enabling seamless name resolution (yourdb.database.windows.net resolves to the private IP). They also work with **service endpoint policies**, allowing fine-grained control over which subnets can access which storage accounts.

A key architectural consideration is managing the **scalability and visibility** of Private Endpoints. For example, if multiple applications require access to the same storage account from different VNets, you'll need to create multiple Private Endpoints. Network design should account for Private Link Service quotas, DNS resolution strategy, and cross-VNet access control policies.

Provision Virtual Networks

Establish the virtual networks vnet-hub and vnet-spoke utilizing Azure Bicep.

vnet-hub.bicep

Overview

Defines the hub virtual network (vnet-hub) with address space and a subnet for Azure Firewall.

```
param location string

resource vnet_hub 'Microsoft.Network/virtualNetworks@2021-05-01' = {
```

```
  name: 'vnet-hub'
  location: location
  properties: {
    addressSpace: {
      addressPrefixes: [
        '10.0.0.0/16'
      ]
    }
    subnets: [
      {
        name: 'AzureFirewallSubnet'
        properties: {
          addressPrefix: '10.0.1.0/24'
        }
      }
    ]
  }
}
output vnetName string = vnet_hub.name
```

vnet-spoke.bicep

Defines the spoke virtual network (vnet-spoke) with address space and a subnet for AKS workloads.

```
param location string

param aksSubnetName string = 'aks-subnet'

param aksSubnetPrefix string = '10.1.1.0/24'

resource vnet_spoke 'Microsoft.Network/virtualNetworks@2021-05-01' = {
  name: 'vnet-spoke'
  location: location
  properties: {
```

```
    addressSpace: {
      addressPrefixes: [
        '10.1.0.0/16'
      ]
    }
    subnets: [
      {
        name: aksSubnetName

        properties: {
          addressPrefix: aksSubnetPrefix
        }
      }
    ]
  }
}

output vnetName string = vnet_spoke.name
```

Create a Private Endpoint for Azure Storage

   1. Private Endpoint

```
        resource privateEndpoint 'Microsoft.Network/
        privateEndpoints@2021-08-01' = {
          name: 'pe-storage'

          location: location

          properties: {
            subnet: {
              id: subnet.id
            }

            privateLinkServiceConnections: [
              {
                name: 'peconn-storage'

                properties: {
                  privateLinkServiceId: storageAccount.id
```

```
          groupIds: [
            'blob'
          ]
        }
      }
    ]
  }
}
```

2. Private DNS Zone

   ```
   resource privateDnsZone 'Microsoft.Network/
   privateDnsZones@2020-06-01' = {
     name: 'privatelink.blob.${environment().suffixes.storage}'

     location: 'global'
   }
   ```

3. Virtual Network Link

   ```
   resource privateDnsLink 'Microsoft.Network/privateDnsZones/
   virtualNetworkLinks@2020-06-01' = {
     name: 'link-vnet-core'

     parent: privateDnsZone

     properties: {
       virtualNetwork: {
         id: virtualNetwork.id
       }
       registrationEnabled: false
     }
   }
   ```

4. DNS Zone Group

   ```
   resource dnsZoneGroup 'Microsoft.Network/privateEndpoints/
   dnsZoneGroups@2021-08-01' = {
     name: 'default'
   ```

```
    parent: privateEndpoint
    properties: {
      privateDnsZoneConfigs: [
        {
          name: 'privatelinkBlobZone'
          properties: {
            privateDnsZoneId: privateDnsZone.id
          }
        }
      ]
    }
  }
```

Real-World Use Case: FinTech Microservices in a Secure Mesh

A FinTech platform offers digital banking and loan services through a distributed microservices architecture. The platform comprises multiple AKS clusters, databases (SQL and Cosmos DB), and event ingestion pipelines. All services are containerized and deployed across three Azure regions for regulatory and latency purposes.

The architecture uses

- A **hub-and-spoke** network design where shared services (Azure Firewall, VPN Gateway, Bastion) reside in the hub VNet

- **Spoke VNets** for AKS clusters, each with its own subnet for system nodes, user nodes, and application gateways

- **Private Endpoints** for Azure SQL and Key Vault, eliminating public access and enforcing Zero Trust

- **Custom route tables** that direct outbound traffic from app subnets through Azure Firewall for inspection and logging

- **Private DNS Zones** linked to each VNet, enabling consistent resolution of private endpoints across the mesh

This topology supports both agility and control. Developers can deploy and scale microservices independently in their spoke VNets, while security and networking policies are centrally managed and enforced in the hub.

In summary, Azure Virtual Networks, subnets, and Private Endpoints are the foundation of a secure and scalable cloud infrastructure. They enable you to segment workloads, control access, and privately consume Azure services, all while maintaining performance and compliance. These constructs are not just network primitives; they are architectural levers for enabling Zero Trust, defense in depth, and regulatory resilience.

In the next section, we'll extend this architecture with advanced **security enforcement mechanisms**, introducing Azure Firewall, DDoS Protection, and Network Security Groups in detail and showing how they form a layered defense system for enterprise-grade deployments.

3.2 Azure Firewall, DDoS Protection, and Network Security Groups (NSGs)

In traditional data centers, network security was often enforced at a central gateway, a perimeter firewall that scrutinized inbound traffic, blocked threats, and enforced access rules. But in cloud-native architectures, where services are distributed, ephemeral, and elastic, the perimeter is fluid. Security must be applied not just at the edge but across every layer of the environment.

Azure addresses this through a **defense-in-depth model**, where multiple security layers from the outer edge to internal subnets protect workloads. This includes **Azure Firewall**, **DDoS Protection**, and **Network Security Groups (NSGs)**, each with a unique role in enforcing secure network boundaries and traffic control.

Azure Firewall: The Cloud-Native, Stateful Packet Inspection Layer

Azure Firewall is a **managed, stateful firewall-as-a-service** that provides Layer 3–7 filtering, threat intelligence, logging, and full integration with Azure Monitor. Unlike NSGs, which are stateless and scoped to NICs or subnets, Azure Firewall operates as a centralized control point ideal for inspecting and controlling traffic between application tiers, virtual networks, or outbound flows to the internet.

Azure Firewall enforces

- **Application Rules**: Filter outbound web traffic based on FQDNs (e.g., only allow access to microsoft.com).

- **Network Rules**: Control traffic based on IP, protocol, and port.

- **Threat Intelligence**: Automatically block known malicious IPs and domains using Microsoft's global threat feed.

- **DNAT/SNAT**: Translate external requests to internal resources or mask internal source IPs.

Architecturally, Azure Firewall is deployed into its own **dedicated subnet (AzureFirewallSubnet)** within a hub VNet. Custom **User-Defined Routes (UDRs)** are applied to force subnet traffic through the firewall. This is known as **forced tunneling** and ensures all egress or inter-VNet traffic is inspected.

Example: In a multitier enterprise application, you may direct all outbound traffic from your app and db subnets to Azure Firewall. Application rules in the firewall allow traffic to specific APIs, SaaS endpoints, or update servers while blocking everything else. Logs are streamed into Log Analytics or Sentinel for real-time detection.

With **Firewall Policy**, you can manage rules centrally and apply them across multiple firewall instances, crucial for enterprises operating in multiple regions or business units.

DDoS Protection: Absorbing Internet-Scale Attacks

Distributed Denial of Service (DDoS) attacks flood applications with overwhelming traffic, rendering services unresponsive. While Azure's infrastructure provides basic protection by default, **Azure DDoS Protection Standard** adds advanced capabilities for enterprise workloads exposed to the internet.

Key features include

- **Adaptive Tuning**: Automatically adjusts thresholds based on your historical traffic patterns.

- **Real-Time Telemetry**: Integrated into Azure Monitor for insights and alerts.

- **Mitigation Policies**: Automatically drops malicious packets during an attack without manual intervention.

- **Cost Protection**: Refunds for scale-out costs incurred during mitigation. Azure DDoS Protection Standard includes a financial safeguard that covers the additional costs incurred when your applications automatically scale out in response to a DDoS attack. If a legitimate attack trigger increased usage such as more compute instances or elevated network throughput, Microsoft will refund the associated scale-out charges, ensuring that customers are not financially penalized for maintaining service availability during an attack.

DDoS Protection is enabled at the **VNet level** and automatically applies to all public IPs within that VNet load balancers, app gateways, VMs, etc. In a public-facing application (e.g., a citizen services portal or ecommerce site), this protection ensures uptime during volumetric attacks without requiring extra infrastructure.

A retail enterprise running a flash sale during a holiday season might see a 20× spike in traffic. DDoS Protection ensures that legitimate users are served while malicious traffic is filtered at the edge before it consumes back-end compute or triggers autoscaling.

Network Security Groups (NSGs): The Access Gatekeepers

NSGs are **stateless, rule-based filters** that control inbound and outbound traffic to individual subnets or NICs. They are the most granular layer of Azure's network security model and are typically used to implement **micro-segmentation** within a VNet.

An NSG rule consists of

- **Priority**: Evaluation order, with lower numbers evaluated first
- **Direction**: Inbound or outbound
- **Protocol**: TCP, UDP, or Any
- **Port Ranges**: Source/destination ports
- **Source/Destination**: IP ranges or service tags (e.g., Internet, VirtualNetwork, AzureLoadBalancer)
- **Action**: Allow or Deny

CHAPTER 3 AZURE NETWORKING AND SECURITY

For example, an NSG applied to a database subnet might allow inbound traffic only from the application subnet on port 1433 (SQL) and deny all other traffic. Similarly, a front-end subnet might allow HTTP/S traffic from the Internet but restrict egress to only approved APIs.

NSGs support **diagnostic logs**, which can be streamed to Log Analytics or Event Hubs for analysis. This is critical for auditing, troubleshooting, and proving compliance in industries like finance or healthcare.

When designing with NSGs, it's crucial to avoid rule sprawl and overlap. Use **naming conventions**, **tagging**, and **automation** (e.g., Terraform, Bicep) to ensure NSGs are consistently applied and version-controlled.

Comparing NSG Rules and Azure Firewall Network Rules

Feature	Network Security Groups (NSGs)	Azure Firewall (Network Rules)
Scope of Application	Subnet or NIC (VM-level)	Centralized at hub VNet (perimeter-level)
Statefulness	**Stateful** (tracks connection states)	**Stateful**
Rule Types	Inbound/outbound access control	Network rules (Layer 3/4) and application rules (Layer 7)
Protocol Support	TCP, UDP, or Any	TCP, UDP, ICMP, and FQDN-based filtering for HTTP/S (via App Rules)
Logging and Analytics	Basic logs via NSG Flow Logs (Network Watcher)	Deep diagnostics, threat intelligence, and traffic analytics
Rule Granularity	Port, protocol, IP ranges/service tags	Fully qualified domains (FQDNs), IP groups, and network-level policies
Recommended Usage	Micro-segmentation, VM, or subnet-level filtering	Centralized security policy enforcement and inter-VNet traffic control
Integration with AVNM	Yes (via security admin rules in Azure Virtual Network Manager)	Yes (part of broader governance models)
Cost	Free	Paid service

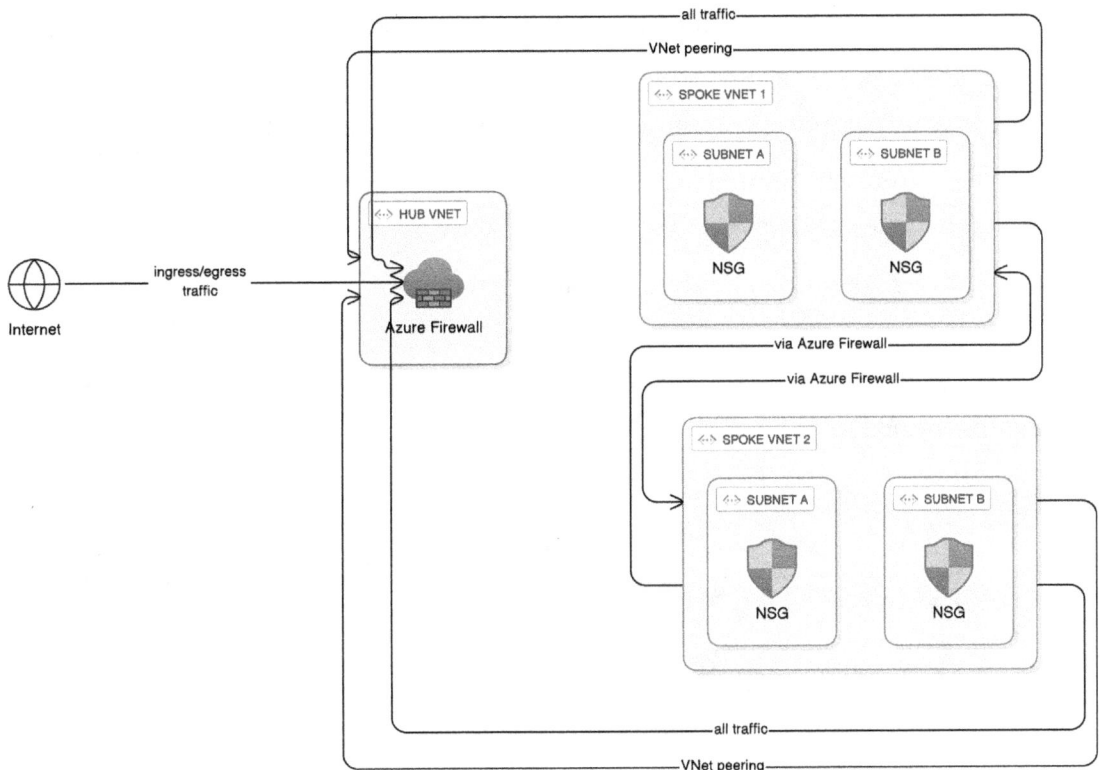

Figure 3-2. *Secure hub-and-spoke network using Azure Firewall for internet ingress, VNet peering, and NSG-protected subnets in spoke VNets*

NSG rules are applied at the **spoke subnet level**, enforcing local access control and micro-segmentation. **Azure Firewall** resides in the **hub VNet**, acting as the centralized perimeter filtering ingress, egress, and inter-spoke traffic through defined network and application rules.

Layered Security in Practice: A Government Services Portal

Consider a government portal that allows citizens to pay taxes, apply for documents, and access services. The architecture includes

- A **public front end** hosted behind Application Gateway with Web Application Firewall (WAF)

- An **Azure Firewall** in the hub VNet, inspecting all east–west and outbound traffic

- **Private Endpoints** for back-end services like Azure SQL and Storage

- **NSGs** on each subnet, restricting lateral movement and enforcing service-specific access

- **DDoS Protection** Standard enabled on the front-end IP

This layered model ensures that even if a malicious actor breaches one layer (e.g., a misconfigured NSG), other controls like the firewall, endpoint isolation, or DDoS mitigation continue to protect the application.

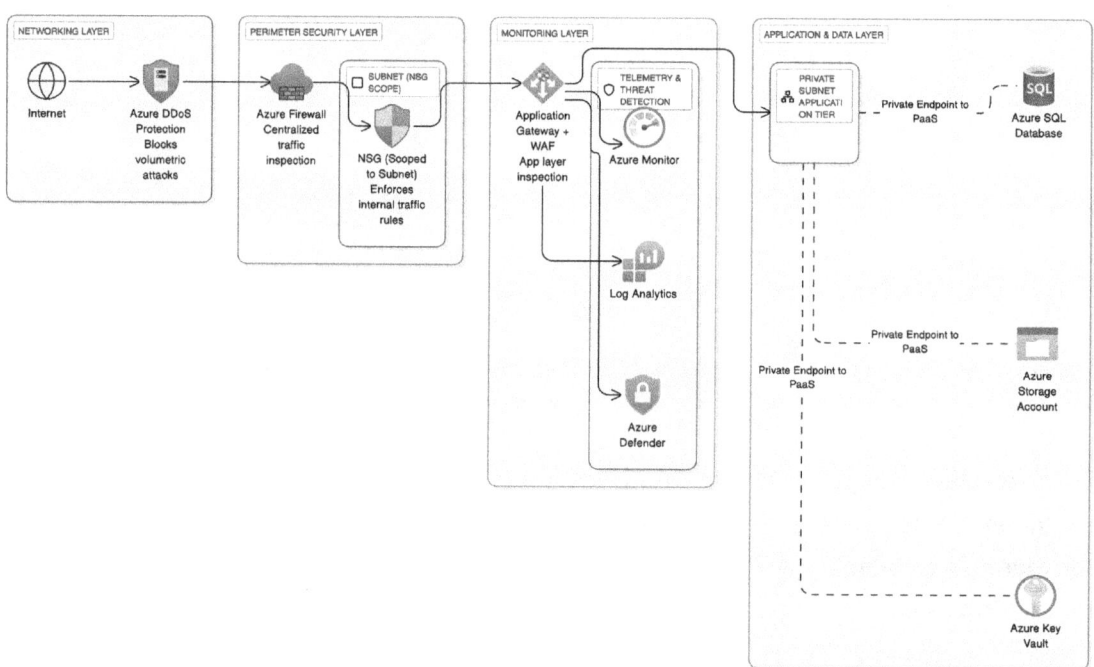

Figure 3-3. *Layered network security architecture showing Networking, Perimeter Security, Monitoring, and Application & Data protections in Azure*

This enhanced diagram illustrates a **Defense-in-Depth security architecture** in Microsoft Azure, designed to align with modern enterprise security requirements, particularly in regulated industries such as finance, healthcare, and government. It demonstrates a layered security model that applies the **Zero Trust principle** "never trust, always verify" from the public internet all the way down to sensitive data access via private PaaS integrations.

Networking Layer

The architecture begins at the **Internet Edge**, where incoming traffic is first intercepted by **Azure DDoS Protection**. This service provides automatic detection and mitigation of Distributed Denial-of-Service attacks by analyzing global traffic patterns and leveraging the Microsoft backbone network. It ensures high availability and operational continuity even during volumetric attacks.

Traffic then passes through **Azure Firewall**, a stateful, centrally managed firewall-as-a-service. It provides Layer 3–7 inspection, application-level filtering, threat intelligence-based filtering, and logging. Azure Firewall enforces security policies based on fully qualified domain names (FQDNs), IP groups, and application rules, offering granular control over both inbound and outbound flows.

Perimeter Security Layer

To complement perimeter protection, the diagram incorporates **Network Security Groups (NSGs)** applied at the **subnet level**. This is a critical enhancement over VM-level NSGs, as subnet-scoped NSGs allow for centralized control of traffic flowing between subnets (east–west) and from external sources (north–south).

These NSGs act like virtual firewalls within the virtual network (VNet), filtering traffic based on IP, protocol, port, and direction. By placing NSGs on the subnet, organizations can segment workloads and enforce stricter boundaries across internal trust zones. For example:

- Only allow traffic from trusted IP ranges.
- Deny all inbound traffic except specific service ports.
- Allow outbound to Azure Monitor or Private DNS zones.

Monitoring Layer

The **Application Gateway** is placed in the next tier, configured with **Web Application Firewall (WAF)** policies to protect against Layer 7 attacks such as SQL injection, XSS, and other threats listed in the OWASP Top 10. This component also handles SSL termination, traffic routing, and session affinity for back-end applications.

Connected to the Application Gateway are integrated monitoring services:

- **Azure Monitor** for telemetry collection and alerts
- **Log Analytics** for querying, visualizing, and correlating logs
- **Microsoft Defender for Cloud** for threat detection and compliance posture

This layer ensures that every activity, policy violation, or anomaly can be observed, audited, and responded to in near real time.

Application and Data Layer

At the lowest tier, the diagram highlights a **Private Subnet** that hosts application workloads. Crucially, access to Azure Platform-as-a-Service (PaaS) offerings such as **Azure SQL Database**, **Azure Storage Account**, and **Azure Key Vault** is enabled via **Private Endpoints**.

Private Endpoints extend the VNet by assigning a private IP address from the subnet to the PaaS resource, eliminating the need for public access. These endpoints

- Use **Azure Private Link** to map traffic securely over Microsoft's backbone network.
- Prevent data exfiltration by avoiding exposure to the public internet.
- Allow fine-grained access policies using Azure RBAC and NSGs.

In the diagram, these connections are represented by **dotted lines** from the private subnet to the respective PaaS services, each labeled clearly to show the secure, internal-only nature of the communication.

This enhanced Defense-in-Depth architecture diagram embodies modern cloud security design for enterprises embracing **Zero Trust**, **segmented networking**, and **secure service integration**. It demonstrates how Azure-native controls, when combined thoughtfully, can create a secure, observable, and governable environment that aligns with compliance frameworks such as **ISO 27001**, **HIPAA**, **PCI-DSS**, and **NIST 800-53**.

By visually emphasizing subnet-level NSGs and Private Endpoint usage, the diagram moves beyond theory to reflect best practices used in real-world production Azure environments.

In summary, Azure provides a comprehensive suite of network security capabilities, each addressing a different tier in the modern threat landscape. Azure Firewall centralizes outbound and inter-subnet inspection. DDoS Protection shields against volumetric attacks. NSGs implement fine-grained access control at the subnet or NIC level. Together, they enable you to build **Zero Trust, policy-driven networks** that resist intrusion, adapt to scale, and meet compliance with confidence.

In the next section, we will explore how these foundational elements align with the **Zero Trust security model,** where identity, telemetry, and continuous verification replace static perimeters as the new control surface of cloud security.

Create and Configure Azure Firewall

firewall.bicep

Deploys Azure Firewall to the hub VNet and links it to AzureFirewallSubnet.

```bicep
param location string

param vnetName string

param firewallSubnetName string

resource publicIp 'Microsoft.Network/publicIPAddresses@2021-05-01' = {
  name: 'fw-public-ip'

  location: location

  sku: {
    name: 'Standard'
  }

  properties: {
    publicIPAllocationMethod: 'Static'
  }
}

resource firewall 'Microsoft.Network/azureFirewalls@2021-05-01' = {
  name: 'fw-prod'

  location: location

  properties: {
```

```
    sku: {
      name: 'AZFW_VNet'
      tier: 'Standard'
    }
    ipConfigurations: [
      {
        name: 'fw-config'
        properties: {
          subnet: {
            id: resourceId('Microsoft.Network/virtualNetworks/subnets',
            vnetName, firewallSubnetName)
          }
          publicIPAddress: {
            id: publicIp.id
          }
        }
      }
    ]
  }
}
```

nsg.bicep

Creates a Network Security Group (NSG) and associates it with the aks-subnet in the spoke VNet.

```
param location string

param nsgName string

param vnetName string

param subnetName string

param addressPrefix string


resource nsg 'Microsoft.Network/networkSecurityGroups@2021-05-01' = {
```

```
name: nsgName

location: location

properties: {
  securityRules: [
    {
      name: 'Allow-HTTP'

      properties: {
        priority: 100

        access: 'Allow'

        direction: 'Inbound'

        protocol: 'Tcp'

        sourcePortRange: '*'

        destinationPortRange: '80'

        sourceAddressPrefix: '*'

        destinationAddressPrefix: '*'
      }
    }
    {
      name: 'Allow-HTTPS'

      properties: {
        priority: 200

        access: 'Allow'

        direction: 'Inbound'

        protocol: 'Tcp'

        sourcePortRange: '*'

        destinationPortRange: '443'
```

```
          sourceAddressPrefix: '*'
          destinationAddressPrefix: '*'
        }
      }
    ]
  }
}
resource subnet 'Microsoft.Network/virtualNetworks/subnets@2021-05-01' = {
  name: '${vnetName}/${subnetName}'
  properties: {
    addressPrefix: addressPrefix
    networkSecurityGroup: {
      id: nsg.id
    }
  }
  dependsOn: [
    nsg
  ]
}
```

3.3 Zero Trust Security Model in Azure

The days of perimeter-based security are behind us. In a traditional on-premises model, trust was assigned based on location: inside the firewall meant safe; outside meant risky. But in the cloud, this assumption is obsolete. Users connect from anywhere. Devices are often unmanaged. Applications span regions and clouds. And threats can originate from both outside and within. In this reality, trust must be **dynamic**, **context-aware**, and **continuously verified**.

This is the essence of **Zero Trust,** a security model that asserts *"never trust, always verify."* Microsoft has embedded Zero Trust into the fabric of Azure, enabling enterprises to enforce security policies based not on static boundaries but on identity, device health, network signals, and behavior. Implementing Zero Trust in Azure involves applying its principles across **identity**, **device**, **network**, **application**, and **data** layers.

Principle 1: Verify Explicitly

Verification begins with **strong identity control. Azure Active Directory (Azure AD)** is the foundational service for **identity, authentication, and authorization** within the Microsoft cloud ecosystem. It governs how users, applications, and devices establish their identities and gain access to resources. Every interaction with Azure services begins with authentication through Azure AD, validating **who** the entity is. Once authenticated, Azure AD enforces **authorization** policies to determine **what** actions the identity is permitted to perform, **on which resources**, **under which conditions**, and **from which context** (such as device compliance, location, or risk level).

This unified approach enables conditional access, role-based access control (RBAC), and policy enforcement to operate seamlessly across Azure, Microsoft 365, and third-party applications. As a result, Azure AD not only proves identity but also ensures that **only the right people or services have the right access to the right resources at the right time**, aligning with modern Zero Trust principles.

Key components include

- **Multi-factor Authentication (MFA)**: Requires users to provide a second verification factor, such as a phone app or biometric input. Enforced using Conditional Access policies.

- **Conditional Access**: Policies that define access rules based on user role, device compliance, location, risk level, or application sensitivity. For example, block access to the Azure Portal for users outside approved IP ranges or on noncompliant devices.

- **Privileged Identity Management (PIM)**: Manages and audits Just-in-Time access for high-privilege roles. Ensures admin rights are not permanent and require approval.

In a healthcare environment, for instance, a radiologist accessing a clinical imaging application may require MFA only when logging in from outside the hospital network, while access from a managed workstation inside the hospital can proceed seamlessly.

Principle 2: Use Least Privilege Access

Zero Trust enforces the **principle of least privilege** across all layers. Users and services are granted only the minimum permissions required, and access is time-bound, just-in-time, and just-enough.

This is achieved through

- **Role-Based Access Control (RBAC)**: Defines what actions an identity can perform on Azure resources. Roles like Reader, Contributor, or custom-defined scopes can be assigned to users, groups, or service principals.

- **Managed Identities**: Allow Azure services (e.g., Functions, VMs, AKS) to authenticate to other Azure services without storing secrets. Identities are automatically managed and rotated.

- **Just-In-Time VM Access**: Through Microsoft Defender for Cloud, RDP and SSH access to VMs can be time-locked, reducing the attack surface.

For example, an automation pipeline that deploys infrastructure only needs Microsoft.Resources/deployments/write access not global contributor rights. Overprovisioning leads to lateral movement risks if credentials are compromised.

Principle 3: Assume Breach

Zero Trust assumes that every access request may be malicious. As such, detection, segmentation, and response mechanisms must be baked into the design.

Effective network-level segmentation is foundational to a defense-in-depth strategy in Azure. It not only isolates workloads but also minimizes lateral movement and exposure by enforcing granular traffic controls across virtual networks. Key components of this segmentation model include

- **Subnets with Network Security Groups (NSGs):** Workloads are placed into discrete subnets, each governed by NSGs that define explicit inbound and outbound traffic rules. This allows administrators to establish trust boundaries between application tiers, development environments, or tenant workloads.

- **Private Endpoints:** These enable secure, private access to Azure PaaS services such as Azure SQL, Storage, and Key Vault over the Microsoft backbone, effectively bypassing the public internet and mitigating the risk of data exfiltration.

- **Azure Firewall and Web Application Firewall (WAF):** Azure Firewall provides stateful packet inspection, threat intelligence filtering, and traffic logging, while WAF enforces Layer 7 rules to defend against OWASP Top 10 threats. Together, they deliver application-aware perimeter security.

- **Azure Bastion:** By enabling secure, browser-based RDP and SSH access to virtual machines over SSL, Azure Bastion eliminates the need to expose management ports like 3389 or 22 to the internet. This reduces the attack surface associated with brute-force login attempts.

- **Default Blocking of High-Risk Ports:** As a best practice, **high-risk management ports such as RDP (3389), SSH (22), Telnet (23), and SMB (445)** should be **explicitly denied by default** using NSGs or Azure Firewall. These ports are frequent targets for reconnaissance and brute-force attacks and should only be opened selectively under tightly controlled access scenarios, preferably using Just-in-Time (JIT) access mechanisms.

Monitoring and detection are enforced with

- **Microsoft Defender for Cloud:** Continuous assessment and threat detection across Azure resources

- **Sentinel (SIEM):** Correlates logs from Azure, on-prem, and third-party sources to surface anomalies

- **Log Analytics and Azure Monitor:** Collect metrics and logs for proactive alerting and visualization

Let's take a real-world scenario. In a digital bank running in Azure, customer service reps access CRM dashboards, while back-end APIs handle sensitive transactions. Using Conditional Access, the CRM access is allowed only during business hours from managed endpoints. Back-end services run in isolated subnets, protected by NSGs and Azure Firewall, and monitored continuously by Defender. Even if an account is compromised, lateral movement is blocked, and alerts are triggered before damage spreads.

Data Security and Labeling

Zero Trust doesn't stop at access; it extends to data. Azure Information Protection and Microsoft Purview allow classification and labeling of sensitive information, so that data policies follow the content, not just the container.

For instance, a document labeled "Confidential - Finance" cannot be downloaded to unmanaged devices or emailed externally, even if the user has read access to the SharePoint site. Similarly, encryption-at-rest and in-transit are enforced across all Azure storage services, with options for customer-managed keys (CMK) stored securely in Azure Key Vault. To further enhance security, **TLS/SSL certificates used for securing communications and authenticating services must be configured for automatic rotation**, typically every **90 days**, to reduce the risk of compromise due to expired or exposed credentials. This principle applies equally to service principal secrets, managed identity certificates, and application-level certificates managed through Azure Key Vault or App Service Certificates.

Zero Trust is not a product. It's a philosophy, and in Azure, it is implemented through a combination of identity services, network segmentation, continuous monitoring, and least privilege enforcement. When applied holistically, it transforms cloud environments from trusted zones into **verified pathways** where every access is validated, every resource is protected, and every anomaly is detected.

In the final section of this chapter, we'll explore how Azure supports **hybrid cloud connectivity**, extending secure networks beyond the cloud into on-premises data centers, enabling consistent policy enforcement, and completing the enterprise cloud security model.

CHAPTER 3   AZURE NETWORKING AND SECURITY

Enable Just-in-Time (JIT) VM Access with Azure Security Center

```
az security jit-policy create
--resource-group rg-secure
--location eastus
--vm /subscriptions/.../resourceGroups/rg-secure/providers/Microsoft.Compute/virtualMachines/my-vm
--name jit-access-policy
--ports '[{"number":22,"protocol":"","allowedSourceAddressPrefix":"","maxRequestAccessDuration":"PT1H"}]'
```

Enforcing Conditional Access with Microsoft Graph PowerShell SDK

Conditional Access is a cornerstone of modern identity security, enabling adaptive policy enforcement based on user roles, device posture, and risk levels. Historically, administrators implemented these policies using the AzureAD or AzureADPreview PowerShell modules. However, these modules are now deprecated in favor of the Microsoft Graph PowerShell SDK, which provides a unified and extensible platform for managing identity and access controls across Microsoft 365 services.

The example below illustrates how to define a conditional access policy that enforces Multi-factor Authentication (MFA) for users in the **Company Administrator** role when a **high sign-in risk** is detected. This implementation leverages the Microsoft Graph SDK and requires the Policy.ReadWrite.ConditionalAccess permission scope.

First, connect to Microsoft Graph with the appropriate scope:

```
Connect-MgGraph -Scopes "Policy.ReadWrite.ConditionalAccess"
```

Then, define and create the conditional access policy using the New-MgIdentityConditionalAccessPolicy cmdlet:

```
New-MgIdentityConditionalAccessPolicy -DisplayName "Require MFA for Admins" `
  -State "enabled" `
  -Conditions @{
    Users = @{
```

```
            IncludeRoles = @("62e90394-69f5-4237-9190-012177145e10")
            # Company Administrator role template ID
        }
        SignInRiskLevels = @("high")
    } `
    -GrantControls @{
        Operator = "or"
        BuiltInControls = @("mfa")
    }
```

In this configuration:

- The **IncludeRoles** field references the **GUID** of the *Company Administrator* role. Microsoft Graph requires role template IDs instead of human-readable names. For example, the role ID 62e90394-69f5-4237-9190-012177145e10 corresponds to the Global Administrator role.

- The **SignInRiskLevels** field targets only sign-ins that Microsoft Entra ID flags as high risk, based on behavioral analytics and threat intelligence.

- The **GrantControls** block specifies that MFA must be satisfied when the conditions are met, using a logical **OR** operator to allow for future extensibility if additional controls are added.

It is important to note that the Microsoft Graph SDK is the forward-looking interface for identity governance. Transitioning to Graph-based policy management ensures compatibility with future features, API versioning, and cross-service integration, an essential step for any organization standardizing on Azure AD as its central identity provider.

Before running this script, ensure that the Graph SDK is installed and up-to-date:

```
Install-Module Microsoft.Graph -Scope CurrentUser -Force
```

This approach exemplifies best practices for implementing **Zero Trust** principles in enterprise environments, particularly in scenarios where administrative privileges require higher scrutiny and stricter authentication controls.

3.4 Implementing Hybrid Cloud Networking

For many enterprises, cloud adoption is not an all-or-nothing decision; it's a continuum. Some workloads remain on-premises due to legacy constraints, compliance requirements, or data gravity, while others shift to the cloud for agility, scalability, or innovation. This hybrid reality demands secure, seamless connectivity between on-premises data centers and Azure environments. Azure meets this challenge through a suite of networking services designed for enterprise-grade hybrid architectures: **VPN Gateway**, **ExpressRoute**, and **Azure Arc**.

Implementing hybrid networking is not simply about connectivity; it's about **preserving trust boundaries**, **extending governance**, and **enabling workload portability** across cloud and edge environments.

VPN Gateway: Encrypted Connectivity Over Public Infrastructure

Azure VPN Gateway enables secure connectivity between an Azure Virtual Network and on-premises infrastructure by establishing **IPsec/IKE encrypted tunnels** over the public internet. These tunnels use the **Internet Protocol Security (IPsec)** suite in combination with the **Internet Key Exchange (IKE)** protocol to create a **mutually authenticated, encrypted communication channel** between two endpoints. IPsec ensures data confidentiality and integrity by encrypting packets at the network layer, while IKE is responsible for negotiating cryptographic keys and security associations used to establish and maintain the secure tunnel. Together, they form the backbone of secure **site-to-site VPN connections**, allowing organizations to safely extend their on-premises network into Azure without exposing data to interception or tampering in transit. It supports:

- Policy-based and route-based VPNs
- Active-standby and active-active configurations for high availability
- Bandwidth up to 10 Gbps with VPN Gateway SKU configurations
- BGP (Border Gateway Protocol) support for dynamic route propagation

This model is ideal for smaller enterprises or branch offices needing a cost-effective solution. It also serves well for **development and testing scenarios**, where rapid setup is preferred over guaranteed performance.

For example, a regional logistics company may use VPN Gateway to connect its warehouse ERP systems to Azure-hosted analytics services. Even with moderate bandwidth needs, the tunnel is encrypted, governed by NSGs and route tables, and monitored via Azure Monitor logs.

While easy to deploy, VPN Gateway depends on the public internet and is susceptible to latency variations or ISP disruptions. For mission-critical workloads, **ExpressRoute** offers a more deterministic alternative.

ExpressRoute: Private, Dedicated Connectivity for Enterprise-Grade Performance

Azure ExpressRoute provides a high-throughput, low-latency, and private connection between on-premises infrastructure and Azure services by bypassing the public internet entirely. This is achieved through dedicated circuits established via a partner exchange provider, co-location facility, or through **ExpressRoute Direct**. Organizations adopting ExpressRoute benefit from

- **Provisioned bandwidth ranging from 50 Mbps to 100 Gbps**
- **Support for BGP (Border Gateway Protocol)** to enable dynamic route advertisement and policy-based routing
- **Dual-circuit architecture for built-in redundancy and high availability**
- **SLA-backed connectivity** with predictable performance and uptime guarantees
- **Private peering to Microsoft services**, including Azure PaaS (e.g., Azure SQL, Storage), SaaS offerings such as Microsoft 365 and Dynamics 365, and cross-region connectivity via ExpressRoute Global Reach

While ExpressRoute offers significant advantages for latency-sensitive, high-volume, or compliance-bound workloads, it is important to note that this is a **premium connectivity option**. Unlike VPN gateways that operate over the public internet, ExpressRoute requires **dedicated infrastructure** and often involves **recurring costs** associated with carrier provisioning, port fees, and bandwidth commitments.

As such, organizations must carefully assess whether the **workload profiles, compliance requirements, and performance objectives** justify the investment. Mission-critical systems such as real-time transaction processing in banking, healthcare data ingestion, or manufacturing telemetry pipelines often merit ExpressRoute due to their stringent service-level expectations. For other scenarios, such as development, testing, or general business use, a secure site-to-site VPN may provide a more cost-effective alternative.

Evaluating Hybrid Connectivity: VPN Gateway vs. ExpressRoute

Choosing the appropriate hybrid connectivity method is a critical architectural decision that directly impacts the performance, cost, security, and manageability of an enterprise's cloud footprint. Azure provides two primary mechanisms for establishing hybrid connectivity between on-premises environments and the Azure cloud: **Azure VPN Gateway** and **Azure ExpressRoute**.

While both enable organizations to extend their data centers into Azure, they differ fundamentally in their connectivity models, reliability guarantees, and ideal use cases. The following sections provide a narrative comparison and selection guidance based on workload characteristics and enterprise requirements.

Azure VPN Gateway: Internet-Based Secure Connectivity

Azure VPN Gateway creates secure tunnels over the public internet using **IPsec/IKE protocols**, providing encrypted communication between an Azure Virtual Network and an on-premises VPN device or network appliance. It is relatively simple to set up, cost-effective for most usage patterns, and suitable for a wide range of use cases, including development environments, low-risk production systems, or backup connectivity paths.

Because VPN traffic traverses the public internet, performance characteristics such as latency and jitter may vary depending on global internet conditions. Azure VPN Gateway supports aggregate throughput up to 10 Gbps (subject to SKU and aggregation), and while it offers redundancy options and supports BGP for route management, it does not come with the same SLA guarantees as ExpressRoute.

Azure ExpressRoute: Private, SLA-Backed Connectivity

Azure ExpressRoute provides **dedicated, private network connectivity** between on-premises infrastructure and Azure, entirely bypassing the public internet. This is typically achieved through a connectivity provider, a co-location facility, or **ExpressRoute Direct**, which connects customers directly to Microsoft's backbone network.

ExpressRoute supports **bandwidths ranging from 50 Mbps to 100 Gbps**, making it ideal for data-intensive, latency-sensitive, or compliance-critical workloads. It integrates with Azure's Software Defined Network (SDN) via private peering and can also provide access to Microsoft SaaS services such as Microsoft 365 and Dynamics 365 via Microsoft peering.

One of ExpressRoute's most powerful features is **ExpressRoute Global Reach**, which allows organizations to interconnect multiple geographically distributed on-premises sites through Microsoft's network. This can dramatically simplify WAN topology and reduce dependence on traditional MPLS or leased lines.

However, these capabilities come at a cost. ExpressRoute involves ongoing charges for circuit provisioning, bandwidth tiers, and provider interconnects, making it a premium connectivity offering. Organizations must carefully assess whether their workload profiles warrant the investment.

Architectural Comparison Overview

The key differences between VPN Gateway and ExpressRoute are summarized in the table below. This is intended to aid architects in understanding where each solution fits within an enterprise network design.

Dimension	Azure VPN Gateway	Azure ExpressRoute
Connectivity Type	Encrypted tunnel over public internet	Private, dedicated line via carrier or co-location
Typical Use Case	General-purpose, dev/test, or backup connectivity	Critical workloads with SLA and performance demands
Throughput	Up to 10 Gbps (based on SKU and aggregation)	50 Mbps to 100 Gbps (based on provisioning)
Latency	Variable (internet-dependent)	Predictable and consistent
Availability SLA	Limited (dependent on internet conditions)	99.95% with dual circuit redundancy
Encryption	Encrypted (IPsec/IKE)	Private path; encryption optional
Routing Support	Basic BGP (limited) or static routes	Full BGP support with policy-based control
Microsoft Services Access	Azure-only	Azure, Microsoft 365, and Dynamics 365 (via Microsoft peering)
Backup Compatibility	Often used as an ExpressRoute fallback path	Can pair with VPN for redundancy
Global WAN Support	Not supported	Supported via ExpressRoute Global Reach
Cost Profile	Low to moderate (pay-as-you-go)	Premium (port fee + bandwidth + provider costs)

Decision Criteria: When to Choose What

The selection between Azure VPN Gateway and ExpressRoute is often governed by a set of technical and business drivers. The following decision matrix helps align connectivity options with organizational priorities:

Use Azure VPN Gateway if:

- You need quick-to-deploy hybrid connectivity for development or testing environments.
- Your workloads are tolerant to variable latency and internet-based routing.

- Cost containment is a primary concern.
- The connection serves as a backup to an existing ExpressRoute circuit.

Use Azure ExpressRoute if:

- Your workloads are mission-critical and demand low latency with high availability.
- You are dealing with sensitive data or fall under compliance regimes that require traffic isolation.
- You need consistent bandwidth for large data transfers, machine learning pipelines, or real-time applications.
- You need to connect multiple geographic on-premises locations with minimal WAN complexity.
- You require private access to Microsoft SaaS offerings such as Microsoft 365 or Dynamics 365.

Strategic Recommendation

Hybrid connectivity is not one-size-fits-all. Instead, architects should assess the role of each workload in the enterprise portfolio and match it to the appropriate connectivity tier. In many cases, the optimal strategy is **a combination of both**: using **ExpressRoute** for critical production paths and **VPN Gateway** as a **cost-effective fallback or for noncritical environments**.

This layered approach ensures **resilience, performance, and cost-efficiency**, aligning with enterprise architecture principles such as **risk mitigation**, **business continuity**, and **optimized resource allocation**.

ExpressRoute is favored by **financial institutions, healthcare providers, manufacturing firms**, and any organization needing **low latency**, **high throughput**, and **stringent compliance**. For instance, a global bank operating in multiple countries might route internal data ingestion and machine learning pipelines over ExpressRoute circuits that enforce data residency, minimize packet loss, and ensure 99.9% uptime.

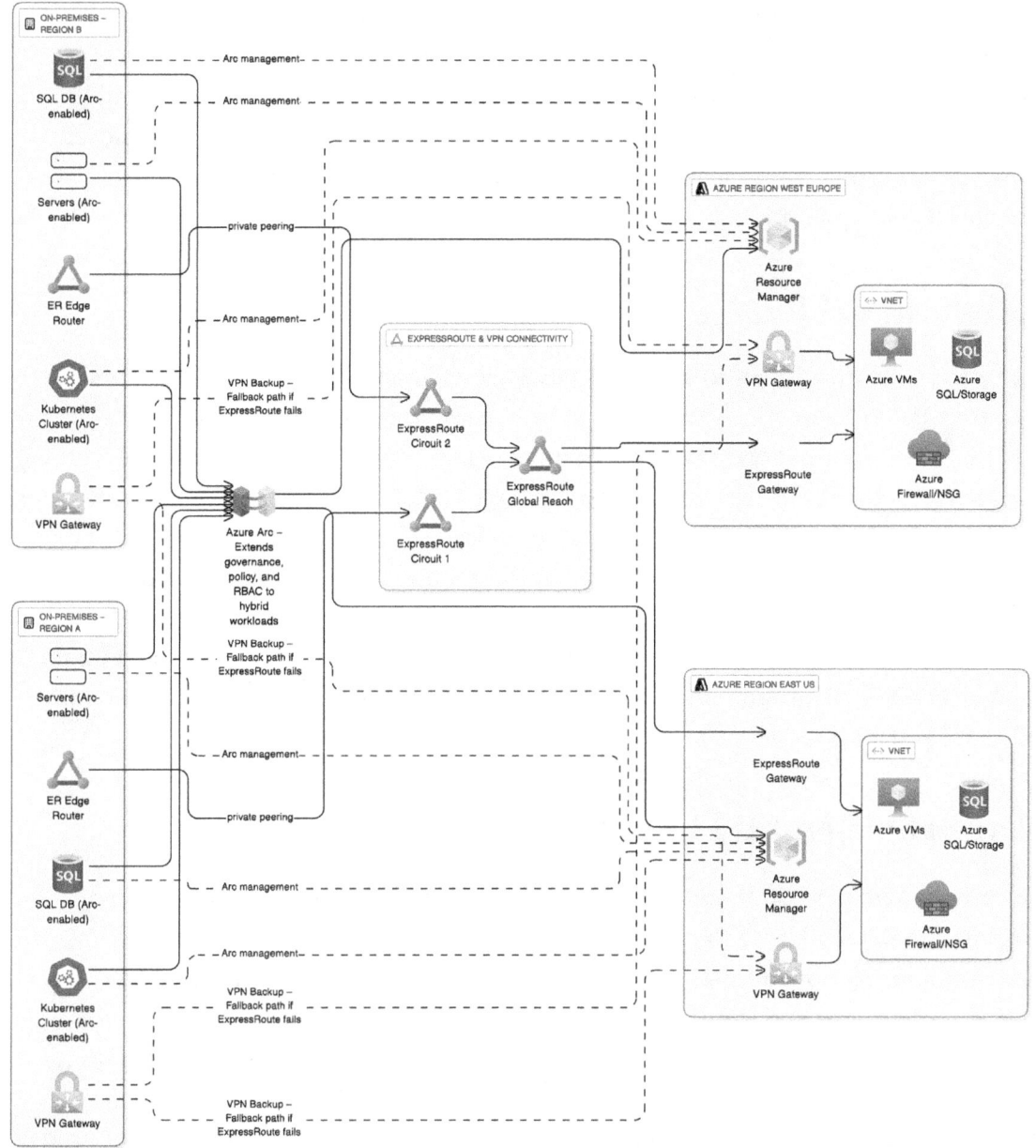

Figure 3-4. *Hybrid cloud network architecture using Azure Arc, ExpressRoute, and VPN backup paths to connect on-premises regions with Azure regions in West Europe and East US, ensuring governance, private peering, and resilient connectivity*

With **ExpressRoute Global Reach**, enterprises can also connect geographically separate on-premises locations through Microsoft's backbone, reducing WAN complexity. Figure 3-4 illustrates a dual-region ExpressRoute setup with private peering and VPN failover for high availability.

This enhanced diagram illustrates a **Defense-in-Depth security architecture** in Microsoft Azure, designed to align with modern enterprise security requirements, particularly in regulated industries such as finance, healthcare, and government. It demonstrates a layered security model that applies the **Zero Trust principle** "never trust, always verify" from the public internet all the way down to sensitive data access via private PaaS integrations.

Networking Layer

The architecture begins at the **Internet Edge**, where incoming traffic is first intercepted by **Azure DDoS Protection**. This service provides automatic detection and mitigation of Distributed Denial-of-Service attacks by analyzing global traffic patterns and leveraging the Microsoft backbone network. It ensures high availability and operational continuity even during volumetric attacks.

Traffic then passes through **Azure Firewall**, a stateful, centrally managed firewall-as-a-service. It offers Layer 3 and 7 inspection, application and threat intelligence filtering, and logging. Azure Firewall enforces security policies based on fully qualified domain names (FQDNs), IP groups, and application rules, offering granular control over both inbound and outbound flows.

Perimeter Security Layer

To complement perimeter protection, the diagram incorporates **Network Security Groups (NSGs)** applied at the **subnet level**. This is a critical enhancement over VM-level NSGs, as subnet-scoped NSGs allow for centralized control of traffic flowing between subnets (east–west) and from external sources (north–south).

These NSGs act like virtual firewalls within the Virtual Network (VNet), filtering traffic based on IP, protocol, port, and direction. By placing NSGs on the subnet, organizations can segment workloads and enforce stricter boundaries across internal trust zones. For example:

- Only allow traffic from trusted IP ranges.
- Deny all inbound traffic except specific service ports.
- Allow outbound to Azure Monitor or Private DNS zones.

Monitoring Layer

The **Application Gateway** is placed in the next tier, configured with **Web Application Firewall (WAF)** policies to protect against Layer 7 attacks such as SQL injection, XSS, and other threats listed in the OWASP Top 10. This component also handles SSL termination, traffic routing, and session affinity for back-end applications.

Connected to the Application Gateway are integrated monitoring services:

- **Azure Monitor** for telemetry collection and alerts
- **Log Analytics** for querying, visualizing, and correlating logs
- **Microsoft Defender for Cloud** for threat detection and compliance posture

This layer ensures that every activity, policy violation, or anomaly can be observed, audited, and responded to in near real time.

Application and Data Layer

At the lowest tier, the diagram highlights a **Private Subnet** that hosts application workloads. Crucially, access to Azure Platform-as-a-Service (PaaS) offerings such as **Azure SQL Database**, **Azure Storage Account**, and **Azure Key Vault** is enabled via **Private Endpoints**.

Private Endpoints extend the VNet by assigning a private IP address from the subnet to the PaaS resource, eliminating the need for public access. These endpoints:

- Use **Azure Private Link** to map traffic securely over Microsoft's backbone network.
- Prevent data exfiltration by avoiding exposure to the public internet.
- Allow fine-grained access policies using Azure RBAC and NSGs.

In the diagram, these connections are represented by **dotted lines** from the private subnet to the respective PaaS services, each labeled clearly to show the secure, internal-only nature of the communication.

This enhanced Defense-in-Depth architecture diagram embodies modern cloud security design for enterprises embracing **Zero Trust**, **segmented networking**, and **secure service integration**. It demonstrates how Azure-native controls, when combined thoughtfully, can create a secure, observable, and governable environment that aligns with compliance frameworks such as **ISO 27001**, **HIPAA**, **PCI-DSS**, and **NIST 800-53**.

By visually emphasizing subnet-level NSGs and Private Endpoint usage, the diagram moves beyond theory to reflect best practices used in real-world production Azure environments.

Extending Governance and Identity with Azure Arc

Hybrid networking is not just about packets; it's about **governance and policy parity**. Azure Arc extends Azure management and control plane capabilities to on-premises servers, Kubernetes clusters, and databases. This means

- **Azure Policy** can enforce tagging, update compliance, and configuration drift across both cloud and on-prem assets.
- **Defender for Servers** and **Monitor agents** work on physical and virtual machines hosted anywhere.
- **Azure RBAC** and Key Vault can be extended to workloads running outside Azure.

For example, a hospital system with multiple clinics may host critical patient systems on local servers but still manage patch compliance and telemetry via Azure Arc. A manufacturing plant may run Azure Arc–enabled Kubernetes at the edge, where latency is critical, but manage policy enforcement and GitOps pipelines from Azure DevOps.

Arc enables a **unified control plane** allowing organizations to treat on-premises and cloud resources uniformly, applying the same security, monitoring, and governance policies regardless of location.

Routing, DNS, and Shared Services

Hybrid environments often involve **custom routing**. User-Defined Routes (UDRs) are used to steer traffic through firewalls, VPN appliances, or on-premises proxies. **Azure Route Server** allows dynamic exchange of routes between your **Network Virtual Appliance (NVA)** such as a third-party firewall or router deployed in a hub network and Azure's Software Defined Network (SDN), improving route management in complex topologies.

DNS architecture must also be carefully planned. Azure Private DNS Zones can be linked across VNets and integrated with on-premises DNS via **conditional forwarding** or **Azure DNS Private Resolver**. This ensures consistent name resolution across hybrid services, which is critical when using **Private Endpoints**, **hybrid AKS clusters**, or **multitier architectures**.

Finally, **shared services** such as Active Directory Domain Services, patch management, and certificate authorities can be hosted in the **hub network** and accessed from both Azure and on-premises workloads.

Deploying an Azure Virtual Network Gateway with Bicep

In hybrid cloud architectures where secure connectivity is required between on-premises infrastructure and Azure, the **Virtual Network Gateway** plays a foundational role. This component facilitates encrypted communication over **IPsec/IKE VPN tunnels**, forming the Azure endpoint for **site-to-site** or **point-to-site** connections.

The following Bicep template automates the deployment of a **Route-Based VPN Gateway** in Azure. It uses modular parameters to promote reusability and environment-specific customization.

```
@description('Name of the Virtual Network Gateway')
param gatewayName string
@description('Azure region')
param location string
@description('Resource ID of the GatewaySubnet')
param gatewaySubnetId string
@description('Resource ID of the Public IP Address')
param publicIpId string

resource vnetGateway 'Microsoft.Network/
virtualNetworkGateways@2023-02-01' = {
  name: gatewayName
  location: location
  properties: {
    ipConfigurations: [
      {
        name: 'vnetGatewayConfig'
        properties: {
          subnet: {
            id: gatewaySubnetId
          }
          publicIPAddress: {
```

```
          id: publicIpId
        }
      }
    }
  ]
  gatewayType: 'Vpn'
  vpnType: 'RouteBased'
  enableBgp: false
  sku: {
    name: 'VpnGw1'
  }
  vpnGatewayGeneration: 'Generation1'
}
dependsOn: [] // Explicit dependsOn not needed as resourceId parameters
are used
}

output gatewayId string = vnetGateway.id
output gatewayIp string = vnetGateway.properties.ipConfigurations[0].
properties.publicIPAddress.id
```

In summary, implementing hybrid networking in Azure is about more than connecting data centers; it's about creating a **cohesive, secure, and governable fabric** that spans physical and virtual boundaries. Whether using VPN Gateway for flexibility, ExpressRoute for performance, or Azure Arc for control plane unification, the goal is the same: to support modern workloads with agility, resilience, and visibility wherever they reside.

With hybrid connectivity in place, we complete the network and security foundation for enterprise-ready Azure deployments. In the next chapter, we move into **high availability and disaster recovery**, exploring how to design resilient architectures that survive failure, recover gracefully, and ensure business continuity across zones and regions.

3.5 Leveraging Azure Virtual Network Manager for Consistent Network Security

As enterprise cloud environments grow in complexity, the management of network resources and enforcement of security policies across virtual networks becomes increasingly fragmented. Traditional controls such as Network Security Groups (NSGs), route tables, and Azure Firewalls are powerful in isolation, but they were never designed to operate as a unified control plane across sprawling environments. For organizations managing hundreds of VNets across different subscriptions or business units, maintaining consistency, compliance, and security posture becomes a daunting operational challenge.

Azure Virtual Network Manager (AVNM) was introduced to address precisely this problem. It acts as a **centralized orchestration layer for network groupings and policy enforcement**, enabling cloud architects to define, apply, and monitor connectivity and security rules at scale. Unlike NSGs, which are scoped to subnets or NICs, AVNM works across regions and subscriptions and can be applied at the **management group level**, bringing governance into alignment with enterprise cloud strategy.

At its core, AVNM enables two key constructs:

- **Network Groups**, which allow logical grouping of virtual networks based on tags, naming conventions, or resource hierarchies.

- **Configuration Policies**, which define either connectivity (hub-and-spoke, mesh) or security (custom rulesets similar to NSGs) that can be automatically deployed across all VNets in a group.

This model allows enterprises to abstract away the manual, error-prone process of attaching NSGs to individual subnets or configuring peering connections one by one. Instead, a cloud administrator can create a global rule, for example, "deny all inbound traffic from the Internet except via Azure Firewall," and enforce it across hundreds of VNets simultaneously, regardless of subscription boundaries.

From a practical standpoint, the **benefits of integrating AVNM at the management group level** include

- **Centralized Network Governance**: AVNM provides a "single pane of glass" to visualize and control virtual networks across your organization. Network groups can reflect organizational structures such as business units, environments (dev, test, prod), or geographical regions, aligning technical controls with business context.

- **Uniform Security Posture**: By defining global security admin rules, enterprises can eliminate policy drift between environments. For example, if a policy mandates that no subnet should be publicly accessible without going through Azure Firewall, AVNM ensures this is enforced uniformly and cannot be bypassed by local NSG misconfigurations.

- **Scalability Across Subscriptions and Regions**: Whether your footprint spans 3 subscriptions or 30, AVNM scales the application of connectivity and security policies without requiring you to replicate infrastructure code or scripts across environments.

- **Policy-Driven Compliance and Auditing**: Because AVNM integrates with Azure Policy and Azure Resource Graph, organizations can audit network configurations in real time and prove compliance with internal or external security baselines such as NIST, ISO 27001, or CIS.

To illustrate its impact, consider a global financial services provider with regulated operations in multiple countries. Each regional office runs a set of VNets for line-of-business applications, yet corporate policy mandates that all ingress traffic must flow through a centralized inspection point. Without AVNM, this would require maintaining a sprawling mesh of NSGs, route tables, and peering connections, each prone to configuration drift. With AVNM, the architecture team defines a security configuration once, applies it to all regional network groups, and monitors enforcement centrally.

In this way, Azure Virtual Network Manager represents a shift from **manual configuration to policy-as-governance**, enabling cloud infrastructure teams to move faster without compromising security or operational discipline. It is especially powerful in **hub-and-spoke architectures**, where consistency of traffic flow enforcement is paramount.

As a best practice, AVNM should be introduced alongside your **management group and landing zone strategy**, ensuring that new environments automatically inherit the correct network posture from the outset. This proactive governance model lays the foundation for Zero Trust networking principles and supports the broader goals of a secure, scalable, and compliant Azure enterprise footprint.

3.6 Summary

In this chapter, we dissected the backbone of any Azure infrastructure: networking. The cloud is inherently a distributed system, and without deliberate, secure, and scalable networking design, the entire architecture becomes brittle and vulnerable. We began by establishing the role of **Azure Virtual Networks (VNets)** as the core building blocks that emulate on-premises networking constructs in the cloud, offering complete IP control, subnet segmentation, and private routing.

We explored how **subnet architecture** and **Network Security Groups (NSGs)** together shape the internal trust boundaries within VNets. With real-world analogies, we illustrated how east–west and north–south traffic flows are governed and why micro-segmentation – paired with **Application Security Groups (ASGs)** – helps simplify security rule management in microservices or role-based deployments.

A key takeaway in this chapter was the distinction between **traditional perimeter defense** and **modern defense-in-depth**, where security is layered from the DNS and routing level down to workload identity and encryption. We highlighted how Azure's **DDoS Protection**, **Azure Firewall**, and **Web Application Firewall (WAF)** provide multiple layers of proactive defense, while **Private Endpoints** and **Service Endpoints** ensure that platform services like Azure SQL or Azure Storage remain accessible only through private IP space shielded from public internet threats.

A significant portion of the chapter focused on **hybrid connectivity**, particularly through **VPN Gateway** and **Azure ExpressRoute**. We clarified when each is appropriate and how to design for enterprise-scale requirements, including BGP route propagation, high availability through active-active configurations, and ExpressRoute circuits with private peering.

Incorporating **Azure Virtual Network Manager (AVNM)**, we extended the discussion into centralized policy enforcement. AVNM allows enterprises to define and apply connectivity and security group rules consistently across VNets and regions. We examined its significance in at-scale environments where governance, repeatability, and automation are paramount.

In the next chapter, we explore how to architect for **high availability and disaster recovery (HA/DR)** in Azure, ensuring business continuity, fault tolerance, and resilience across regions, zones, and service tiers.

CHAPTER 4

High Availability and Disaster Recovery

Designing for Resilience, Redundancy, and Business Continuity in Azure

Failure in the cloud is not a question of "if"; it is a question of "when." From transient disk I/O latency to full-scale regional outages, failure in distributed systems is not only inevitable, but it is also often unpredictable. In response, enterprise architecture has matured from merely optimizing for uptime to actively engineering for **resilience**. This shift is not just technical; it is strategic. Availability and recoverability now shape board-level decisions, influence regulatory compliance, and determine business continuity in a globally connected economy.

Microsoft Azure provides a rich ecosystem of services for building resilient systems, ones that can withstand hardware failures, datacenter disruptions, and regional disasters. But using these services effectively requires more than just enabling a check box for redundancy. True cloud resilience demands a layered design strategy: from local fault tolerance using Availability Zones and scale sets to regional disaster recovery using Azure Site Recovery and from snapshot-based backup policies to active-active multi-region topologies that keep your systems operational even during a catastrophe.

This chapter lays out that strategy in structured, actionable form. We begin by examining the architectural foundation of high availability, constructing virtual machine environments that can tolerate component-level failures. Then, we explore how to automate disaster recovery using native Azure services, implement intelligent backup strategies, and ultimately, build globally resilient systems that can fail gracefully while maintaining state, trust, and service guarantees.

4.1 Designing for High Availability (VM Scale Sets, Load Balancers)

At its core, **high availability (HA)** is about ensuring that services remain accessible and performant despite failures within a specific fault domain, whether that be a failed virtual machine, a downed physical rack, or a datacenter experiencing infrastructure disruption. Azure provides two fundamental mechanisms for designing HA systems at the infrastructure level: **redundant compute distribution** and **load-balanced traffic management**.

The most foundational unit of compute availability in Azure is the **Availability Set**. An Availability Set is a logical grouping that instructs Azure to distribute your virtual machines across multiple physical servers, compute racks, storage units, and network switches. This distribution spans what Azure terms **fault domains** (different hardware) and **update domains** (staggered OS or platform maintenance windows). For workloads that cannot yet be containerized or modernized such as stateful legacy applications or third-party enterprise software, Availability Sets provide the minimum safeguard against infrastructure-based disruptions.

However, Availability Sets are confined to a single datacenter within a region. For higher levels of redundancy and fault tolerance, Azure offers **Availability Zones**. These are physically separated locations within an Azure region, each with independent power, cooling, and networking. By deploying resources across multiple zones, architects can create zone-resilient solutions that are insulated from even large-scale infrastructure events. Services like Virtual Machine Scale Sets, Azure Load Balancer, and Azure Kubernetes Service all natively support zone-aware deployments.

The modern enterprise workload, however, rarely consists of a single VM. It is composed of multiple loosely coupled services that scale based on demand. **Virtual Machine Scale Sets (VMSS)** are designed precisely for this scenario. A VMSS defines a template for your application nodes and automatically scales the number of instances up or down in response to load, scheduled business hours, or predictive scaling algorithms. VMSS instances are evenly spread across zones when zone balancing is enabled, thereby not only enabling elasticity but also enforcing cross-zone resilience.

Deploying a VM Scale Set Using Azure Bicep

```
@description('Location for VMSS resources')
param location string
```

```
@description('Virtual Network ID')
param vnetId string

@description('Subnet ID for VMSS')
param subnetId string

@description('VMSS name')
param vmssName string

@description('Admin username for VMSS')
param adminUsername string

@description('Admin password for VMSS')
@secure()
param adminPassword string

@description('Instance count for VMSS')
param instanceCount int = 2

@description('VM size for VMSS')
param vmSize string = 'Standard_DS1_v2'

// Load Balancer
resource loadBalancer 'Microsoft.Network/loadBalancers@2023-06-01' = {
  name: '${vmssName}-lb'
  location: location
  properties: {
    frontendIPConfigurations: [
      {
        name: 'LoadBalancerFrontend'
        properties: {
          privateIPAllocationMethod: 'Dynamic'
          subnet: {
            id: subnetId
          }
        }
      }
    ]
```

```
      backendAddressPools: [
        {
          name: 'BackendPool'
        }
      ]
      loadBalancingRules: [
        {
          name: 'HTTPRule'
          properties: {
            frontendIPConfiguration: {
              id: resourceId('Microsoft.Network/loadBalancers/
              frontendIPConfigurations', '${vmssName}-lb',
              'LoadBalancerFrontend')
            }
            backendAddressPool: {
              id: resourceId('Microsoft.Network/loadBalancers/
              backendAddressPools', '${vmssName}-lb', 'BackendPool')
            }
            protocol: 'Tcp'
            frontendPort: 80
            backendPort: 80
            enableFloatingIP: false
            idleTimeoutInMinutes: 4
          }
        }
      ]
    }
}

// Virtual Machine Scale Set
resource vmss 'Microsoft.Compute/virtualMachineScaleSets@2023-03-01' = {
  name: vmssName
  location: location
```

```
properties: {
  upgradePolicy: {
    mode: 'Manual'
  }
  virtualMachineProfile: {
    networkProfile: {
      networkInterfaceConfigurations: [
        {
          name: 'VMSSNIC'
          properties: {
            primary: true
            ipConfigurations: [
              {
                name: 'IPConfig'
                properties: {
                  subnet: {
                    id: subnetId
                  }
                  loadBalancerBackendAddressPools: [
                    {
                      id: resourceId('Microsoft.Network/loadBalancers/
                      backendAddressPools', '${vmssName}-lb',
                      'BackendPool')
                    }
                  ]
                }
              }
            ]
          }
        }
      ]
    }
  }
}
```

```
      osProfile: {
        computerNamePrefix: vmssName
        adminUsername: adminUsername
        adminPassword: adminPassword
      }
      storageProfile: {
        imageReference: {
          publisher: 'Canonical'
          offer: 'UbuntuServer'
          sku: '18.04-LTS'
          version: 'latest'
        }
        osDisk: {
          createOption: 'FromImage'
          managedDisk: {
            storageAccountType: 'Standard_LRS'
          }
        }
      }
    }
    overprovision: true
    singlePlacementGroup: true
    sku: {
      name: vmSize
      capacity: instanceCount
    }
  }
}
output vmssName string = vmss.name
output loadBalancerName string = loadBalancer.name
output backendPoolId string = resourceId('Microsoft.Network/loadBalancers/
backendAddressPools', '${vmssName}-lb', 'BackendPool')
```

For example, an ecommerce platform might deploy its stateless web front ends as a VMSS across three zones within the East US region. During Black Friday or Diwali sales, traffic can spike from thousands to millions of concurrent users. The scale set, monitored via CPU and request queue length, automatically adds nodes to handle increased demand. A **Standard Load Balancer**, positioned in front of the scale set, uses TCP-based health probes to distribute traffic only to healthy instances, removing failed ones from the pool in real time.

Azure Load Balancers serve as the cornerstone of high availability for compute-bound services. At Layer 4 (TCP/UDP), they support both **internal load balancing** for east–west traffic across tiers (e.g., web-to-app) and **public load balancing** for north–south ingress from users. Configurable health probes testing endpoints like /healthz direct traffic to functional instances, preventing partial or degraded service.

Moreover, modern HA designs often complement load balancers with **Application Gateway** or **Azure Front Door** for advanced Layer 7 routing, SSL termination, and application firewalling. These are not just enhancements; they are essential for managing user experience under failure conditions, such as redirecting traffic to healthy regions or serving degraded content from CDN caches.

CHAPTER 4 HIGH AVAILABILITY AND DISASTER RECOVERY

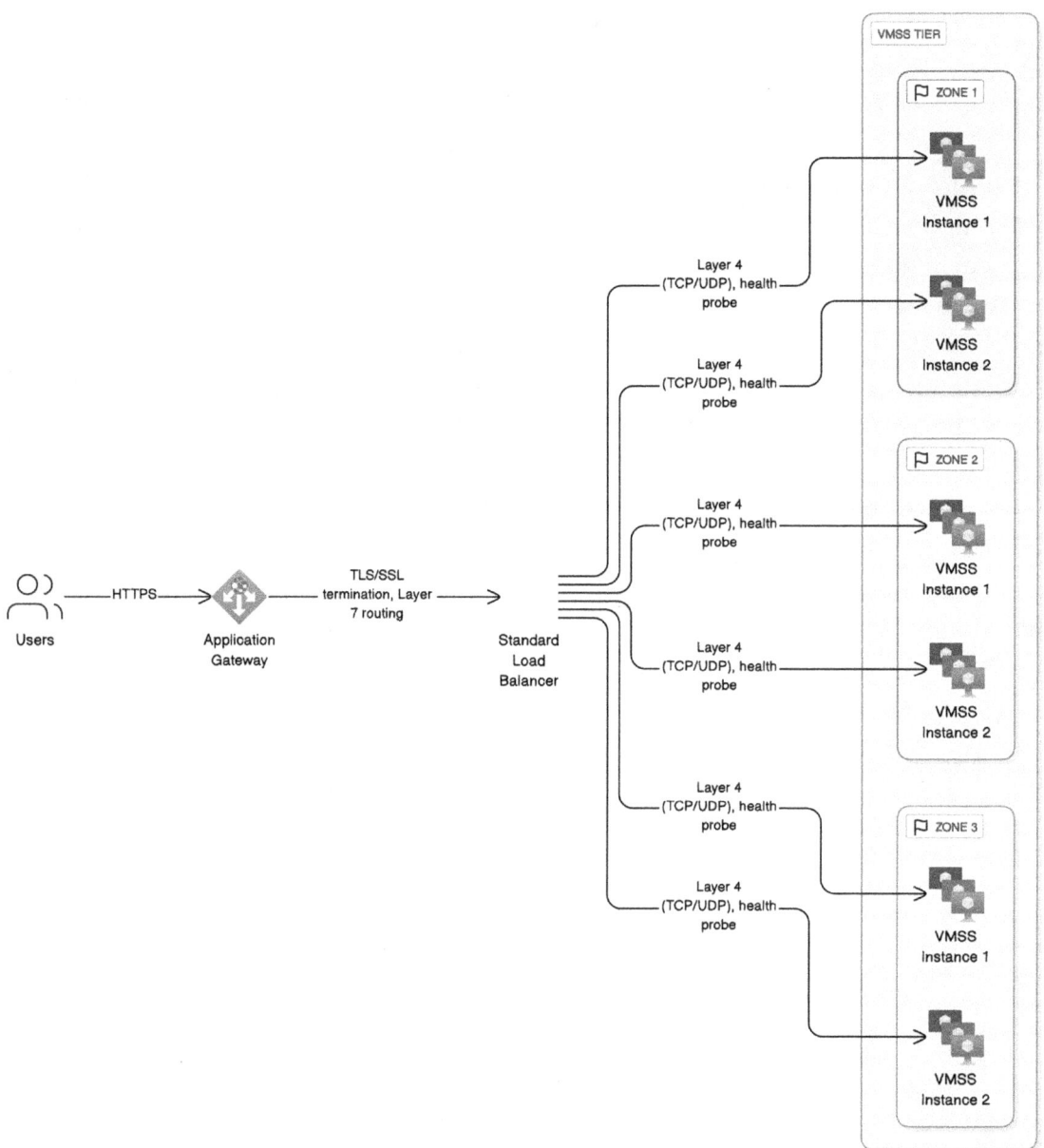

Figure 4-1. *High-Availability (HA) Architecture*

Figure 4-1 illustrates a typical HA topology: VM Scale Sets spanning three Availability Zones, fronted by a Standard Load Balancer, with Azure Application Gateway handling TLS termination and routing to back-end pools.

Inbound traffic from users is routed through a **Standard Azure Load Balancer**, which performs Layer 4 (TCP/UDP) distribution based on real-time health probes. This load balancer directs traffic only to healthy VM instances across the zones, ensuring availability even when certain nodes become unresponsive. Sitting in front of the load balancer is the **Azure Application Gateway**, which operates at Layer 7 to manage HTTP/HTTPS traffic. It provides **TLS termination, cookie-based session affinity**, and **path-based routing**, enhancing user experience and reducing load on back-end nodes.

The VMSS instances are configured for **automatic scaling**, adjusting the number of VMs based on defined metrics such as CPU utilization or HTTP queue depth. This enables the environment to scale dynamically during peak usage such as during seasonal sales or product launches while minimizing costs during periods of lower demand.

This HA design combines horizontal scaling, fault domain awareness, and multi-zone deployment to meet enterprise-grade availability SLAs. It is ideal for mission-critical applications that require both elasticity and robust protection against infrastructure failure.

The design principles established at this layer, redundancy, elasticity, and health-based routing, form the infrastructure-level defense against disruption. Yet availability alone is not enough. When entire regions go offline or a long-term state must be preserved, we enter the realm of **disaster recovery** where failover, replication, and orchestration must be engineered with equal precision.

High availability is the foundation of a resilient infrastructure strategy in Azure. Section 4.1 dissected the mechanisms that ensure localized fault tolerance spanning from virtual machine availability sets to multi-zone virtual machine scale sets and from standard load balancers to advanced Layer 7 routing with Application Gateway. Together, these constructs enable systems to self-heal from node failures, maintain user experience under load, and operate seamlessly across fault domains and update domains.

However, high availability alone does not address broader disruptions such as regional outages, data center evacuations, or site-level maintenance windows. These scenarios demand a more comprehensive strategy that extends beyond automatic failover and incorporates stateful workload recovery, orchestrated resumption of services, and continuity of operations across regions.

That strategy is the domain of disaster recovery, where Azure Site Recovery (ASR) becomes the cornerstone of cross-region resilience and recoverability.

4.2 Disaster Recovery with Azure Site Recovery (ASR)

Disaster recovery (DR) is the architectural safety net that catches systems when availability fails. Where high availability shields against localized failures such as individual node crashes or datacenter outages, disaster recovery prepares systems to withstand **regional loss, infrastructure collapse**, or even **geopolitical disruption**. The goal is not merely to maintain uptime but to **restore operability with minimal data loss and downtime**, aligned with the business's Recovery Time Objective (RTO) and Recovery Point Objective (RPO).

Azure Site Recovery (ASR) is Microsoft's native service for orchestrating disaster recovery across Azure regions and between on-premises infrastructure and Azure. It is a fully managed solution that automates the replication of virtual machines, the configuration of failover plans, and the execution of recovery operations during an outage. Unlike traditional DR approaches, which require a parallel infrastructure and complex manual workflows, ASR embraces **automation, scale, and observability** as core tenets.

Continuous Replication and Recovery Architecture

At the heart of ASR is its **replication engine**, which captures and synchronizes the operating system, data disks, and configuration state of protected machines. This replication is near real-time for Azure-to-Azure DR, with typical RPOs measured in seconds or low single-digit minutes. For on-premises to Azure DR, ASR installs a lightweight Mobility Service agent on each source machine, enabling disk-level replication to a designated storage account in Azure.

These replicated images are held in an unprovisioned state, ready to be instantiated into fully running VMs on demand during failover. This separation of **replication and recovery** allows enterprises to maintain a minimal infrastructure footprint in the DR region until needed, significantly reducing ongoing cost.

CHAPTER 4 HIGH AVAILABILITY AND DISASTER RECOVERY

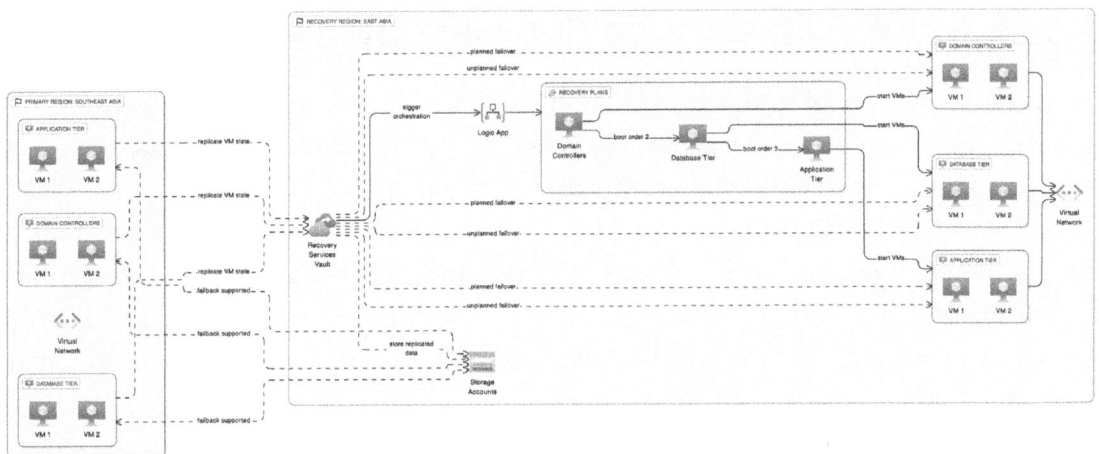

Figure 4-2. *Replication and Failover Flow in Azure Site Recovery*

Figure 4-2 illustrates the replication and failover flow between a primary region (e.g., Southeast Asia) and a paired recovery region (e.g., East Asia), including Recovery Services Vault, target networks, and failover policies.

In the primary region, mission-critical workloads, including domain controllers, application servers, and database tiers, are continuously replicated to the target storage in the secondary region via ASR's near-real-time replication engine. This ensures minimal Recovery Point Objective (RPO) and enables high-fidelity failover readiness.

The diagram shows the distinct phases of disaster recovery: data replication, recovery plan execution, and failover instantiation. Recovery Plans orchestrate tiered boot sequences, ensuring that foundational services such as Active Directory are brought online before dependent applications. Integration with Azure Automation and Logic Apps facilitates automated post-failover tasks, such as DNS updates and endpoint reconfiguration.

Directional arrows indicate both planned and unplanned failover paths, as well as bidirectional data sync for failback after recovery. The architecture is designed to support testing via isolated virtual networks, allowing organizations to validate DR readiness without impacting production systems.

This visual representation underscores the critical role of ASR in enterprise-grade business continuity planning, offering an automated, policy-driven, and secure approach to regional disaster recovery in Azure.

Orchestrated Failover with Recovery Plans

Failover is not just about spinning up compute; it is about restoring services **in the right sequence** and in the **correct topology**. Azure Site Recovery enables this through **Recovery Plans**, which act as declarative workflows for disaster response.

A Recovery Plan can define groups of machines that failover together, specify **boot order dependencies**, and embed **Azure Automation runbooks** or **PowerShell scripts** to perform tasks such as DNS updates, static IP remapping, or service validation. For example, a Recovery Plan for a multi-tier banking application might specify the following sequence:

1. Promote Active Directory domain controllers to ensure identity resolution.
2. Bring up SQL Server databases and restore availability groups.
3. Launch middle-tier APIs after verifying database readiness.
4. Activate front-end application servers, update internal DNS, and publish new endpoints.

This granular control ensures not just recovery but **controlled and validated recovery**.

Moreover, ASR supports **Test Failover,** a feature that allows organizations to simulate disaster recovery in a **sandboxed virtual network**. These test environments mirror production workloads but remain isolated from live traffic, enabling functional verification, security testing, and compliance validation without impact. In sectors like healthcare and finance, where audits demand documented proof of DR preparedness, this capability is indispensable.

Periodic **Test Failovers** are orchestrated quarterly, with compliance reports automatically generated from ASR logs and Azure Monitor workbooks fulfilling the firm's audit trail requirements and demonstrating operational resilience.

Enabling Azure Site Recovery for a VM

```
az backup vault create \
  --name recovery-vault \
  --resource-group rg-dr-site \
  --location westus2
```

```
az backup protection enable-for-vm \
  --vault-name recovery-vault \
  --resource-group rg-dr-site \
  --vm app-vm \
  --policy-name DefaultPolicy
```

Creating a Recovery Plan Using PowerShell

Define a recovery plan with VMs grouped by tiers.

```
$recoveryPlan = New-AzRecoveryServicesAsrRecoveryPlan -Name
"AppTierRecoveryPlan" `
  -PrimaryFabric "AzureEastUS" `
  -RecoveryFabric "AzureWestUS" `
  -RecoveryPlanGroup @(
    @{ GroupName = "DomainControllers"; VMs = @("dc-vm1", "dc-vm2") },
    @{ GroupName = "DatabaseTier"; VMs = @("sql-vm") },
    @{ GroupName = "ApplicationTier"; VMs = @("api-vm", "web-vm") }
  )
```

Failback and Cost Optimization

After an unplanned failover, **failback** must be handled with care. ASR allows workloads running in the recovery region to be resynchronized back to the original site once it becomes available. Failback supports **delta sync**, transferring only changed data, and enables **reverse replication**, making the secondary site the new primary if needed.

ASR is also cost-conscious. Since replicated disks reside in **Azure storage as page blobs** and VMs are not provisioned until failover, the solution incurs minimal compute costs until needed. Organizations can also use **Azure Reserved Instances** or **Spot VM pricing** in their DR region to further optimize expenditure for warm-standby systems.

Security and Policy Integration

Disaster recovery must not compromise security posture. ASR supports integration with **Azure Policy** to ensure that all critical machines identified by resource group, tag, or naming convention are enrolled in replication automatically. It also supports **encryption of replicated data, role-based access control (RBAC)**, and **Network Security Group (NSG)** enforcement on failover VMs.

For highly regulated environments, ASR ensures that replication traffic remains within **private links** or **VPN tunnels**, depending on source and destination configuration. Logs and recovery actions are stored in **immutable storage** where required, satisfying compliance needs for data retention and forensic analysis.

In summary, Azure Site Recovery elevates disaster recovery from a backup plan to a **strategic capability**. It transforms chaotic, high-pressure response scenarios into **repeatable, testable workflows**, aligning technology with business continuity goals. Whether you are protecting a monolithic legacy application or a distributed microservices platform, ASR provides the scaffolding for fast, secure, and validated recovery.

In the next section, we will turn our attention to the **preservation of critical data assets,** examining how Azure Backup, point-in-time restore, and data immutability ensure not only that infrastructure is restorable but that no essential state is ever permanently lost.

4.3 Backup and Data Protection Strategies

While high availability and disaster recovery ensure service continuity, they do not automatically guarantee **data integrity** or **restorability** in the face of corruption, ransomware, human error, or data center compromise. In cloud-native architecture, the separation of compute and storage makes it technically feasible to recreate infrastructure, but if data is lost or compromised, no amount of automation can bring back what never had durable protection.

In Azure, data is not a by-product of services; it is a **first-class asset**, and its protection must be architected with the same rigor as its compute environment. Backup is no longer just a file-level insurance mechanism; it is a **multilayered, policy-driven strategy** that ensures resilience, regulatory compliance, and business continuity. Azure offers a robust set of native services to support these goals, including **Azure Backup, point-in-time restore, immutable blob storage**, and **vault-based data retention**.

Azure Backup: Managed, Policy-Based Protection at Scale

Azure Backup is a fully managed service that protects IaaS virtual machines, Azure Files, SQL Server databases on Azure VMs, and more. It operates through **Recovery Services Vaults**, where all backup data is stored securely with support for **encryption-at-rest**, **geo-redundancy**, and **long-term retention**.

For virtual machines, Azure Backup leverages **application-consistent snapshots**, ensuring that backups include OS-level and workload-level consistency. For example, a backup taken from a running VM with an SQL Server workload will capture the state of the OS, SQL processes, and underlying disk structures in a coordinated snapshot using **Volume Shadow Copy Service (VSS)**.

Backup policies can be tailored to business-specific retention requirements. A typical enterprise policy might specify

- Daily backups retained for 30 days
- Weekly backups retained for 12 weeks
- Monthly backups retained for 1 year
- Yearly backups retained for 7–10 years (for compliance or audit purposes)

This policy-driven model abstracts away infrastructure complexity, ensuring that DevOps and platform teams focus on strategic workload concerns, while the backup system enforces SLA-aligned protection autonomously.

Point-in-Time Restore (PITR) for Azure Databases

For Azure-managed database services, such as **Azure SQL Database**, **Azure PostgreSQL**, and **Azure MySQL**, backups are built-in and invisible to the user but just as powerful. These services support **continuous backups with point-in-time restore**, allowing the database to be rolled back to any moment within the configured retention window (default: 7–35 days).

This capability is critical in environments where

- Accidental deletions or destructive updates must be recoverable.
- Schema corruption or misapplied migrations require rollback.
- Recovery testing must simulate time-based attacks or faults.

For example, if an engineering team accidentally drops a key table at 14:17:23, the platform owner can restore the database to 14:17:00, isolate the restored instance, and export the missing data. This operation can be triggered via the Azure Portal, PowerShell, CLI, or ARM templates, integrating backup into automation pipelines.

Long-term retention (LTR) for Azure SQL is also available, where weekly, monthly, or yearly backups are stored in **cool-tier blob storage** for up to 10 years, satisfying audit and compliance mandates for financial, government, and healthcare sectors.

Backing Up an Azure SQL Database with LTR

```
resource ltrPolicy 'Microsoft.Sql/servers/databases/backupLongTermRetention
Policies@2022-02-01-preview' = {
  name: 'default'
  parent: sqlDatabase
  properties: {
    weeklyRetention: weeklyRetention
    monthlyRetention: monthlyRetention
    yearlyRetention: yearlyRetention
    weekOfYear: 1 // Set the week of the year for yearly retention (e.g., 1
    for the first week)
  }
}
```

Immutable Blob Storage and Object Versioning

Azure Blob Storage serves as the backbone for many critical enterprise workloads ranging from unstructured document archives to data lake analytics. Protecting this data requires more than redundancy; it requires **immutability** and **version control**.

Azure offers several data protection capabilities for blob storage:

- **Soft Delete for Blobs and Containers**: Allows recovery of deleted blobs for a configurable retention period (e.g., 30 days)

- **Blob Versioning**: Automatically maintains previous versions of blobs upon modification

- **Immutable Blob Storage (WORM)**: Enforces write-once-read-many policies using time-based or legal hold retention

These features are indispensable in protecting against ransomware and insider threats. For example, if an attacker overwrites customer contracts stored in blob storage, the system retains all previous versions, each with a unique timestamp, allowing forensic recovery and legal non-repudiation.

In highly regulated environments such as banking, these WORM policies fulfill requirements for **financial statement integrity**, **digital audit trails**, and **regulatory attestation** under standards like SEC 17a-4(f), GDPR, and ISO 27001.

Protecting Containerized State and AKS Workloads

As more enterprises shift toward containerized microservices, traditional backup paradigms become insufficient. Stateless services may not require backup, but **stateful services deployed in Azure Kubernetes Service (AKS)** such as databases, queues, and file-based processing apps must be protected.

Azure integrates with **Velero**, a Kubernetes-native backup tool, to support

- **Persistent volume (PV) snapshotting**

- **Backup and restore of Kubernetes objects (e.g., ConfigMaps, Secrets, Deployments)**

- **Namespace-level rollback and selective recovery**

For example, in a logistics platform running in AKS, a MongoDB StatefulSet backed by Azure Disks may experience corruption due to a bad operator update. Using **Velero**, an open source backup and recovery tool for Kubernetes clusters integrated with Azure Blob Storage, the team can restore both the persistent volume state and the deployment configuration, effectively re-creating the service while preserving transactional data.

This model also supports **multi-cluster DR**, where a backup taken from one cluster (e.g., in West Europe) can be restored to another region (e.g., North Europe), maintaining business continuity in the event of an AKS cluster failure or regional outage.

CHAPTER 4   HIGH AVAILABILITY AND DISASTER RECOVERY

Creating Blob Storage with Soft Delete Using Azure Bicep

```
@description('The name of the storage account.')
param storageAccountName string = 'storagearchiveprod'

@description('The location for the storage account.')
param location string = resourceGroup().location

@description('The SKU for the storage account.')
param skuName string = 'Standard_LRS' // Change to Standard_LRS for compatibility

@description('The retention period for soft delete (in days).')
@minValue(1)
param daysRetained int = 30

@description('The access tier for the StorageV2 account.')
@allowed([
  'Hot'
  'Cool'
])
param accessTier string = 'Hot'

resource storageAccount 'Microsoft.Storage/storageAccounts@2022-09-01' = {
  name: storageAccountName
  location: location
  sku: {
    name: skuName
  }
  kind: 'StorageV2'
  properties: {
    isHnsEnabled: true // Enable hierarchical namespace
    allowBlobPublicAccess: false
    minimumTlsVersion: 'TLS1_2'
    accessTier: accessTier // Add access tier property
  }
}
```

```
resource blobServiceProperties 'Microsoft.Storage/storageAccounts/
blobServices@2022-09-01' = {
  name: 'default'
  parent: storageAccount
  properties: {
    deleteRetentionPolicy: {
      enabled: true
      days: daysRetained
    }
  }
}
```

Governance, Monitoring, and Compliance

Data protection is not only about backing up the right resources; it is about proving that protection is in place, continuously monitored, and aligned with policy. Azure enables this through deep integration with **Azure Policy**, **Azure Monitor**, and **Security Center**.

- **Azure Policy** can enforce that all production VMs, tagged with Environment=Prod, are enrolled in a backup policy within 24 hours of deployment.

- **Azure Monitor** tracks backup success/failure, alerting platform teams if job completion drops below SLA.

- **Defender for Cloud** evaluates backup compliance posture and integrates with regulatory frameworks like HIPAA, NIST, and PCI-DSS.

Backups are also **encrypted by default**, and access to backup vaults is restricted through **role-based access control (RBAC)** and **Private Link endpoints**, ensuring that only authorized automation tools or recovery operators can interact with the vault.

In summary, Azure's backup and data protection ecosystem is broad, mature, and deeply integrated. Whether the need is to restore a single deleted file, recover a mission-critical database to a known-good state, or reconstruct the operational topology of an entire Kubernetes environment, Azure provides the tools and policies to make recovery predictable, secure, and compliant.

With backups securing the integrity of data and disaster recovery plans ensuring restoration pathways, there remains one final layer in the resilience hierarchy: *geographic continuity*. In scenarios where entire Azure regions may go offline due to natural disasters, geopolitical events, or large-scale infrastructure failures, organizations must go beyond isolated recovery and embrace **multi-region deployments** that enable live failover with minimal disruption. This next section explores how to architect such globally resilient systems on Azure.

4.4 Multi-region Deployments and Failover

High availability protects against local failures. Disaster recovery enables service restoration when a disaster strikes. But **multi-region architecture** is what transforms cloud systems from robust to unbreakable. When services must remain uninterrupted even in the face of complete regional outages, geopolitical disruptions, or natural disasters, it is a multi-region design that delivers continuity.

Multi-region deployment in Azure refers to **distributing workloads, data, and user traffic across geographically separate Azure regions**, with failover mechanisms in place to detect outages and route traffic to healthy regions. This architectural approach is the gold standard in **global business continuity,** providing not only resilience but also **latency optimization, regulatory compliance**, and **geo-redundant scale**.

Azure's global infrastructure with 60+ regions and interconnected backbone networks makes multi-region design technically feasible. But to make it enterprise-ready, architects must consider five critical domains: **deployment topology**, **data synchronization, traffic routing, failover orchestration**, and **governance compliance**.

CHAPTER 4 HIGH AVAILABILITY AND DISASTER RECOVERY

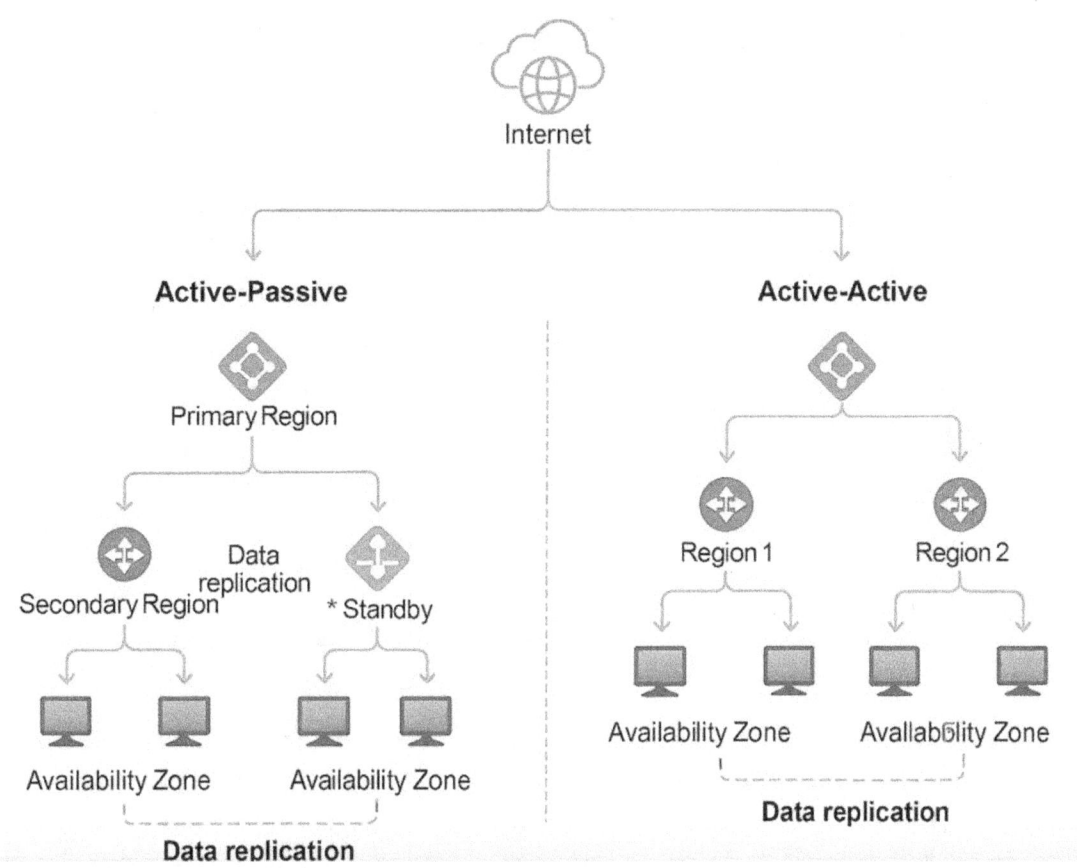

Deployment Topologies: Active-Passive vs. Active-Active

At the heart of multi-region design is the choice between **active-passive** and **active-active** deployment models.

In an **active-passive** setup, the primary region hosts live production traffic, while the secondary (standby) region mirrors the environment in a cold, warm, or hot state. It may replicate data continuously, but compute resources in the standby region are typically underutilized until failover is initiated. This model is more cost-efficient and easier to manage but requires some delay for activation and DNS switchover.

In contrast, an **active-active** design runs live workloads in multiple regions simultaneously. User traffic is distributed based on geography or latency. Data is synchronized in real time, and each region is capable of full workload autonomy. This setup is highly resilient and performant but operationally complex, with challenges around **data consistency**, **conflict resolution**, and **increased cost**.

Data Synchronization and Consistency Models

Ensuring data consistency across regions is one of the most nuanced aspects of multi-region architecture. Azure offers several services with **built-in geo-replication** and **configurable consistency models**:

- **Azure Storage (Blob, Table, Queue)**: Supports **Geo-Redundant Storage (GRS)** and **Read-Access GRS (RA-GRS)**, asynchronously replicating to a secondary region.

- **Azure SQL Database**: Offers **Active Geo-Replication** with readable secondaries and manual failover.

- **Azure Cosmos DB**: Enables **multi-master replication** with tunable consistency from strong to eventual, giving architects control over availability vs. consistency tradeoffs.

- **Azure Database for PostgreSQL and MySQL**: Support read replicas in remote regions for query offloading and backup DR.

Data synchronization strategy must align with the **CAP theorem**: systems can only simultaneously guarantee two of Consistency, Availability, and Partition tolerance. For example, financial systems may require **strong consistency,** favoring a primary region for writes and replicating reads. Ecommerce carts may tolerate **eventual consistency** for performance and user experience.

Architects must also prepare for **split-brain scenarios**, where both regions accept writes but network partitioning prevents synchronization. In such cases, Cosmos DB uses conflict resolution policies based on timestamps or custom logic to determine source-of-truth.

Global Traffic Distribution and Routing

To implement seamless regional failover, Azure provides sophisticated global routing services:

- **Azure Traffic Manager** operates at the DNS level, directing traffic based on endpoint health, performance, geography, or weighted priorities. It is ideal for protocol-agnostic workloads or hybrid deployments involving Azure and on-premises systems.

- **Azure Front Door** operates at Layer 7, offering HTTP/HTTPS routing, global SSL offloading, caching, and integrated Web Application Firewall (WAF). It monitors back-end health and automatically redirects traffic to the next-best region upon detection of failure.

These services support **proactive health probes**, evaluating HTTP status codes, latency, and service response to determine endpoint viability. Routing decisions can be made in milliseconds, and DNS propagation is optimized via low Time-To-Live (TTL) settings for rapid switching.

A multi-region deployment for a healthcare web portal might use Azure Front Door to balance traffic across West Europe and North Europe, with Traffic Manager providing DNS-level redundancy. Application Gateway in each region handles Layer 7 routing and TLS inspection locally, preserving data sovereignty and response time.

Active-Active Multi-region Architecture with Azure Front Door and Cosmos DB

In enterprise-scale applications, especially global, customer-facing platforms like ecommerce, financial trading systems, or travel booking portals, latency and high availability are non-negotiable. Active-active deployment architectures enable services to run concurrently in multiple regions, ensuring fault tolerance and optimal performance for users worldwide.

A robust example of this approach involves pairing **Azure Front Door** with **Azure Cosmos DB's multi-region write (multi-master)** capabilities. Front-end application instances can be deployed across geographically dispersed Azure regions, such as **East US** and **Southeast Asia**, while Cosmos DB automatically replicates data with low-latency consistency between the regions. Azure Front Door, functioning as a Layer 7 global entry point, dynamically routes traffic based on real-time latency metrics, health probes, and geo-based rules.

In the event of a regional service degradation or outage, **Azure Front Door performs automatic failover**, redirecting requests to the healthiest available back end with minimal impact on the end-user experience. This behavior eliminates the need for additional DNS-level tools like Azure Traffic Manager unless non-HTTP(S) services or legacy failover mechanisms are involved.

Below is a high-level architecture diagram to illustrate this deployment pattern.

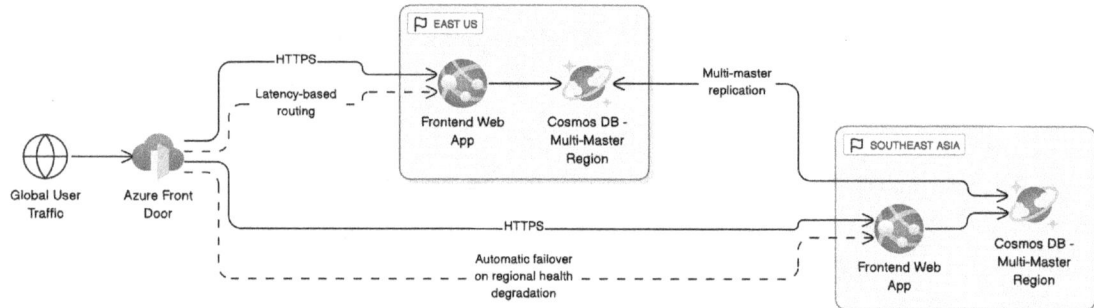

Figure 4-3. *Global traffic management using Azure Front Door with latency routing and multi-region Cosmos DB*

DNS-Level Resilience with Azure Traffic Manager

While Azure Front Door excels at HTTP/HTTPS global routing with built-in TLS termination, path-based routing, and application-layer health probes, there are scenarios where **DNS-level traffic distribution** is better suited, especially for **non-HTTP(S) workloads**, **legacy systems**, or **hybrid deployments** that involve **on-premises endpoints**.

Azure Traffic Manager is a DNS-based traffic load balancer that distributes traffic across Azure regions or external endpoints based on configurable routing methods. Unlike Azure Front Door, which operates at Layer 7, Traffic Manager sits at the DNS layer and resolves client requests to the appropriate IP address or FQDN. Once resolved, the client communicates directly with the selected endpoint.

A typical use case involves active-passive deployments where one region serves traffic while another remains on standby. For example, a financial institution running core services in West Europe with a cold standby in North Europe can use **Traffic Manager with failover routing**. If a health probe detects that the primary endpoint is unhealthy, Traffic Manager automatically resolves DNS queries to the backup region, ensuring business continuity.

CHAPTER 4   HIGH AVAILABILITY AND DISASTER RECOVERY

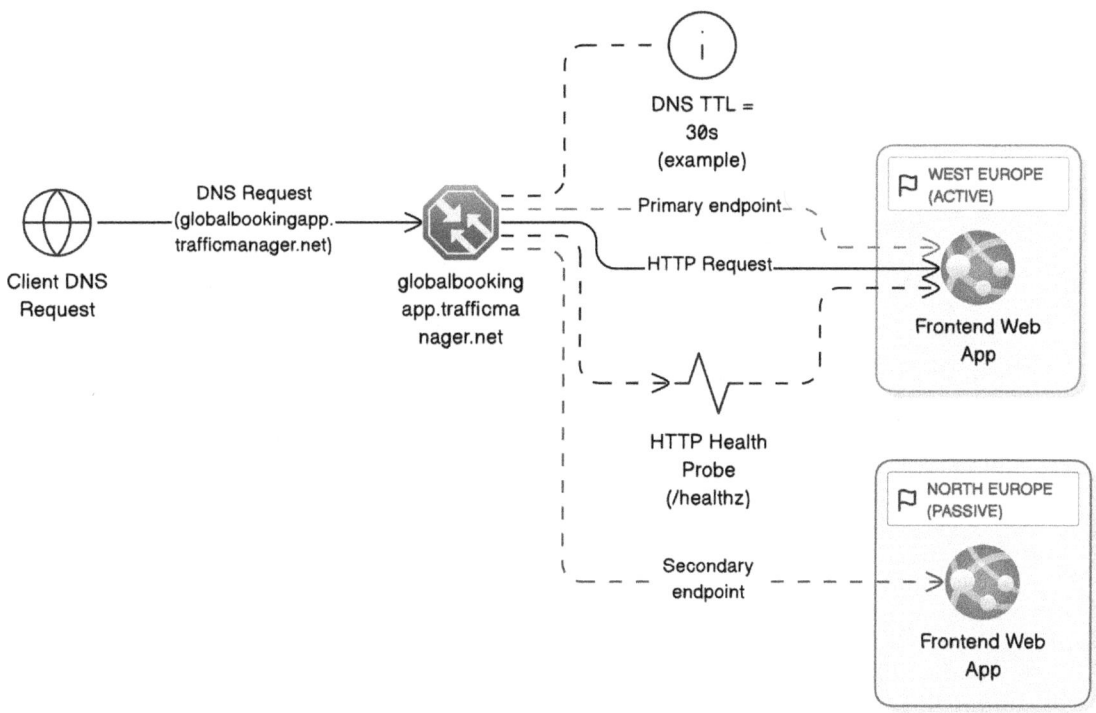

Figure 4-4. *DNS-Level Resilience with Azure Traffic Manager*

Nested Traffic Manager with Azure Service Fabric Across Four Regions

As mission-critical systems demand high throughput, low latency, and geographically distributed resilience, Azure Service Fabric emerges as a powerful platform for building and managing scalable microservices. However, Service Fabric's regional deployments, while robust, need to be combined with intelligent global traffic routing to ensure smooth user experiences and automated failover. This is where **nested Azure Traffic Manager profiles** play a vital role.

In a typical global application architecture, Service Fabric clusters are deployed in **four Azure regions,** for example, **East US, West US, North Europe, and Southeast Asia**. Each cluster hosts identical microservice instances, offering local service affinity and regional high availability. To orchestrate traffic across these clusters, **Traffic Manager is configured in a nested profile structure**, combining both **latency-based routing** and **failover policies**.

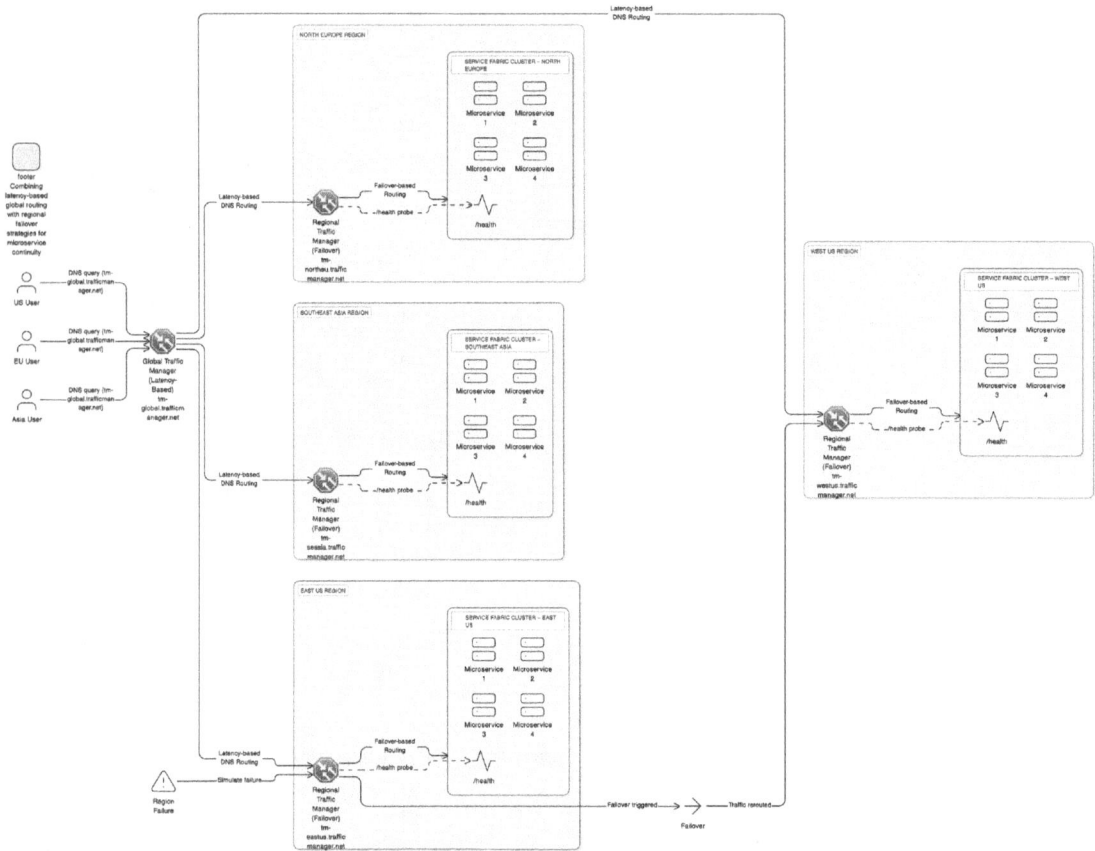

Figure 4-5. Global traffic management with latency-based routing and regional failover via Azure Service Fabric

How Nested Traffic Manager Works

A **parent Traffic Manager profile** routes DNS requests using latency-based routing to the nearest regional endpoint. Each of these regional endpoints is itself a **child Traffic Manager profile** configured with failover or weighted routing among specific nodes or service partitions within that region.

This nested setup offers two advantages:

1. **Optimized Performance:** Users are directed to the lowest-latency region automatically.

2. **Localized Failover:** If a specific Service Fabric node or application endpoint within a region fails, the child profile ensures failover within that region before escalating to a global failover.

When to Use Azure Traffic Manager vs. Azure Front Door

Criteria	Azure Front Door	Azure Traffic Manager
Layer	Application (Layer 7)	DNS (Layer 3/4)
Protocol Support	HTTP/HTTPS only	Any (HTTP, TCP, UDP, custom ports)
Failover Granularity	Sub-second with health probes	DNS TTL-dependent (usually 20–30s)
TLS Termination	Yes	No
Endpoint Support	Azure App Services, AKS, on-prem (HTTP)	Azure VMs, External IPs, On-prem
Ideal Use Cases	Web apps, APIs, microservices	Legacy systems, hybrid, multicloud

Note You can combine Traffic Manager and Azure Front Door when you need both **protocol-agnostic failover** and **smart application-layer routing**, though most modern architectures consolidate into Front Door unless legacy constraints demand DNS-level control.

Orchestration, Automation, and Recovery Readiness

Failover is only as good as its orchestration. Azure provides a spectrum of tools to automate detection, redirection, and recovery:

- **Azure Site Recovery (ASR)** replicates VM and app workloads across regions and orchestrates failover based on predefined recovery plans.

- **Azure Monitor and Log Analytics** detect anomalies in metrics, logs, and service health, triggering alerts or remediation scripts.

- **Azure Automation** and **Logic Apps** can execute workflows such as DNS updates, resource scaling, or service redeployments during failover events.

- **Azure Bicep and Terraform** support region-agnostic IaC templates, enabling infrastructure rehydration in DR regions quickly and consistently.

In highly regulated environments, **runbook testing** is crucial. Enterprises schedule **chaos simulations** to intentionally disable regional components and validate failover response training, SRE teams, and hardening automated playbooks. Azure's built-in support for test failovers ensures that critical services remain within SLA expectations.

Governance, Compliance, and Residency

Beyond technical execution, multi-region deployment introduces questions of **data residency**, **sovereignty**, and **compliance enforcement**. Azure's region-pairing model ensures that backup and recovery data remain in geopolitically aligned pairs, such as UK South UK West or Germany North Germany West Central.

Azure Policy can enforce that sensitive data (e.g., DataClassification=Confidential) is only deployed in regions approved for residency. Combined with **customer-managed keys (CMK)** in Azure Key Vault and **Private Link endpoints**, data movement is both visible and controlled.

Enterprises in healthcare, banking, and government are increasingly required to demonstrate that regional failover does not violate residency laws. Azure provides regional scope audit logs and region-scoped resource locks to support such attestations.

4.5 Navigating the Complexity of On-Prem to Azure Failover Scenarios

Failing over enterprise workloads from an on-premises data center to Azure is far more complex than a straightforward lift-and-shift; it requires navigating a multifaceted set of challenges that encompass network topology, service replication, and platform dependencies.

Service replication, identity integration, and data consistency are critical considerations in designing hybrid failover architectures. While Azure provides a rich portfolio of tools such as Azure Site Recovery (ASR), Azure Arc, and hybrid networking options, orchestrating a seamless failover still demands careful planning and architectural alignment across both environments.

One of the most critical complexities is **replicating stateful services**. Stateless services, such as front-end web apps or APIs, can often be containerized and redeployed to Azure App Service or AKS with relative ease. However, services dependent on persistent data, such as SQL Server databases, Active Directory-integrated applications, or custom services relying on legacy storage back ends, require synchronous or near-real-time replication strategies. For these, Azure Site Recovery offers virtual machine-level replication, but this doesn't automatically translate into application-level consistency or minimal downtime.

Moreover, **network and DNS failover** introduces its own class of intricacy. A common pattern is to use Azure Traffic Manager or Front Door with geo-based routing and health probes, but without a resilient hybrid DNS setup and IP range alignment, services may not resolve or authenticate correctly post-failover. If DNS caching or TTL configurations aren't properly optimized, clients may continue attempting to reach the on-premises endpoint even after it's offline.

Identity synchronization adds another layer of complexity. Azure AD Connect can bridge Azure AD and on-premises Active Directory, but in a failover scenario, loss of domain controller availability can cripple authentication-dependent services. Enterprises often mitigate this by deploying read-write domain controllers in Azure and ensuring line-of-sight to the domain controller infrastructure via ExpressRoute or VPN tunnels.

Finally, **testing and validation** of disaster recovery playbooks are rarely straightforward. Many organizations operate under a false sense of security, assuming that replicating VMs equates to application readiness in Azure. In reality, DR tests must simulate full application failover, including configuration drift, version compatibility, IP addressing, and firewall rules between regions.

In enterprise-grade architectures, embracing **Infrastructure as Code (IaC)**, **pre-provisioned landing zones**, and **policy-based deployments** becomes essential to reduce manual error and accelerate time-to-recovery. Azure Site Recovery, combined with Azure Automation Runbooks and Azure Monitor alerting, can support orchestrated failover workflows, but these must be validated in simulated blackout conditions, ideally on a quarterly basis.

In summary, achieving seamless failover from on-premises to Azure is a nontrivial undertaking. It demands a holistic approach that spans replication strategy, DNS and identity continuity, data consistency, and rigorous testing. Designing for failover from the beginning, not as an afterthought, is key to building production-ready hybrid resilience.

4.6 Summary

In the modern digital-centric enterprise, failure is no longer a remote possibility; it is an inherent architectural certainty. Whether it involves a disk malfunction, a zone-wide network interruption, or a complete regional disaster, systems must be purposefully designed to anticipate, withstand, and recover from such failures. Azure's platform offers extensive capabilities to address these challenges; however, achieving true resilience requires more than simply enabling service features. It necessitates deliberate architectural planning.

We began by dissecting the anatomy of **high availability (HA)**. Azure's native constructs, such as **Availability Sets, Availability Zones,** and **Virtual Machine Scale Sets (VMSS)**, provide the physical and logical separation necessary to withstand faults in compute, network, and power layers. These constructs were paired with **Layer 4 Load Balancers** and **Application Gateways** to distribute traffic across healthy nodes and zones, ensuring continuity of experience even when individual components degrade or fail.

Through CLI-based examples, we translated theory into practice, demonstrating how to deploy scalable, zone-aware VMSS backed by health-aware load balancers. In real-world scenarios like a surge-prone ecommerce platform, we saw how HA patterns mitigate risk and scale elastically under pressure without compromising service integrity.

We then progressed to **disaster recovery (DR)**, the safety net for scenarios where availability zones alone are insufficient. Using **Azure Site Recovery (ASR)**, we explored how replication, orchestration, and automated failover plans transform DR from a manual fail-safe into a continuous, testable, and compliant business function. Recovery Plans, ASR vaults, and scripted failovers ensure that recovery is not just possible but precisely executed in the correct sequence with minimal human intervention and alignment to enterprise RTO/RPO thresholds.

The **backup and data protection** strategies that followed reinforced a key architectural truth: *high availability protects services, but backups protect state.* Through **Azure Backup**, **point-in-time restores (PITR)**, and **immutable storage configurations**, we illustrated how to safeguard business-critical data against logical corruption, operator error, and cyber threats like ransomware. In Kubernetes contexts, we introduced **Velero** as a first-class tool to back up container volumes and cluster state, an emerging best practice in microservices resiliency.

Finally, we elevated the discussion to **multi-region deployments**, the zenith of cloud resilience. Using Azure's global infrastructure, we demonstrated how to architect services that remain operational even in the face of regional blackouts. **Active-passive** and **active-active topologies**, powered by **Azure Traffic Manager**, **Azure Front Door**, and **geo-replicated data services** like Cosmos DB and Azure SQL, allow architects to design applications with *zero dependency on a single location*.

Multi-region design also brought attention to **data consistency models**, **traffic routing heuristics**, and **failover orchestration workflows**. We showed how these strategies are not only technical; they are also legal and regulatory. **Data residency**, **sovereignty**, and **governance** must be architected into every layer, especially in highly regulated industries like finance and healthcare.

Throughout the chapter, layered examples, Bicep/CLI snippets, and production-grade design patterns illustrated how to codify resilience, not just document it. Whether protecting a single VM or a globally distributed SaaS platform, the recurring message was clear: **resiliency is not achieved by configuration; it is earned through design discipline.**

As we transition into the next chapter on **Azure Kubernetes Service (AKS)**, the principles of high availability and disaster recovery will remain foundational. Container orchestration environments add complexity but also unlock new dimensions of scaling, fault isolation, and automation that further strengthen the enterprise's ability to recover quickly and serve reliably.

CHAPTER 5

Azure Kubernetes Service (AKS) for Enterprise Workloads

Orchestrating Containers with Security, Scalability, and DevOps Maturity in Azure

Over the last decade, the software industry has undergone a massive shift from monolithic architectures toward distributed, microservices-based systems. This transformation has been fueled by the rise of containerization, particularly Docker, which enables developers to encapsulate applications and their dependencies into lightweight, portable units that run consistently across environments. But while containers solve the problem of application portability, they introduce a new challenge: orchestrating, scaling, securing, and monitoring hundreds if not thousands of these containers across production-grade environments.

Kubernetes has emerged as the dominant orchestration platform to solve this problem. However, Kubernetes itself is a complex, distributed system with a steep operational learning curve. Managing control planes, performing upgrades, securing workloads, and integrating with networking and identity providers demands expertise and a mature DevOps culture. For enterprises that are adopting Kubernetes but want to avoid the operational burden of managing it, **Azure Kubernetes Service (AKS)** offers a powerful alternative.

AKS is a fully managed Kubernetes service provided by Microsoft Azure. It offloads the responsibility of maintaining the control plane; integrates natively with Azure networking, identity, and security services; and supports advanced features such as autoscaling, DevSecOps, and GitOps. It is not just Kubernetes in the cloud; it is **production-grade Kubernetes optimized for Azure-native environments**.

In this chapter, we explore how AKS becomes the foundation for modern cloud-native enterprise platforms. Each section addresses a critical domain: understanding the architectural building blocks of Kubernetes within AKS, securing clusters to meet Zero Trust standards, implementing dynamic autoscaling and performance optimization, and deploying services using automated pipelines with full GitOps support. The final chapter then connects AKS to the broader observability and governance model, ensuring workloads are not only agile and scalable but also secure, compliant, and cost-aware.

5.1 Introduction to AKS and Kubernetes Fundamentals

Kubernetes, originally developed at Google and now maintained by the Cloud Native Computing Foundation (CNCF), has become the standard for container orchestration in modern enterprise platforms. At its core, Kubernetes provides a distributed control plane that automates the scheduling, deployment, scaling, and healing of containerized applications across clusters of virtual machines. Its appeal lies in a declarative model, in which engineers describe the desired state of the system, and Kubernetes works continuously to make that desired state a reality.

For organizations running on Microsoft Azure, **Azure Kubernetes Service (AKS)** abstracts the operational complexity of managing Kubernetes. It offers a fully managed control plane with built-in integrations to Azure's identity, networking, and monitoring services. This allows cloud architects and platform engineers to focus on building secure, scalable applications without needing to manage cluster internals such as etcd, kube-scheduler, or API server certificates.

To understand the value proposition of AKS, one must begin with the **core components of Kubernetes** and how AKS simplifies or enhances them in the Azure ecosystem.

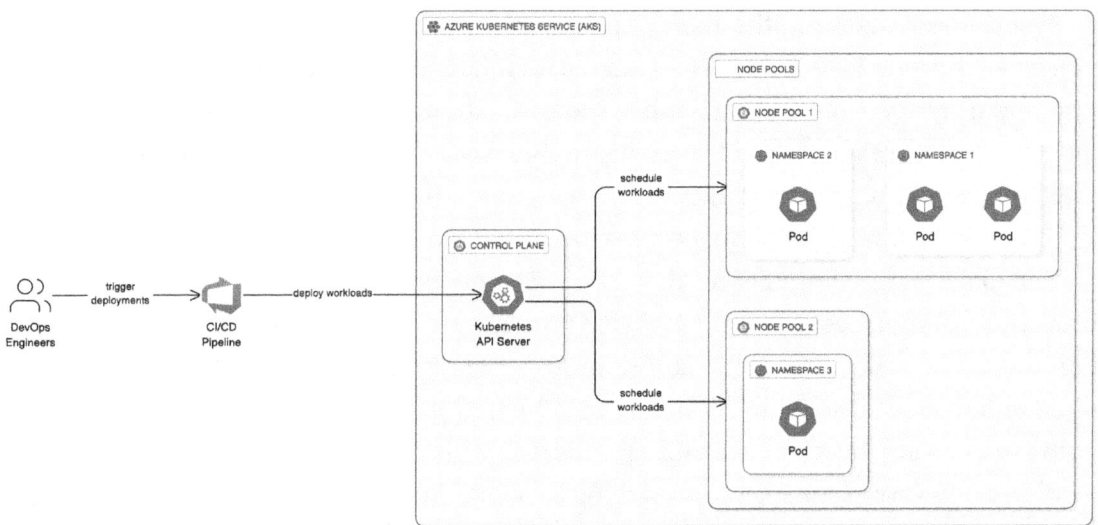

Figure 5-1. *AKS deployment flow from CI/CD pipeline to API Server, control plane, node pools, and namespaces*

This figure illustrates a continuous deployment pipeline for workloads targeting an Azure Kubernetes Service (AKS) cluster. The architecture emphasizes the role of DevOps engineers, CI/CD automation, the Kubernetes control plane, and a flexible, multi-node pool layout utilizing namespaces for workload isolation.

The process begins with DevOps engineers initiating deployments through a CI/CD pipeline, typically implemented using platforms like Azure DevOps, GitHub Actions, or Jenkins. Upon triggering, the CI/CD system packages the workload and interacts with the Kubernetes API Server, the central interface of the AKS control plane. This API server validates and processes deployment manifests and configuration files, including resource definitions like Deployments, Services, ConfigMaps, and Secrets.

The Kubernetes API server is responsible for scheduling workloads across the cluster. In this architecture, the AKS cluster is configured with multiple node pools, each optimized for specific workload types or operational characteristics such as compute-optimized, memory-optimized, or GPU-enabled nodes.

Node Pool 1 hosts two Kubernetes namespaces:

- Namespace 1 contains two application pods, representing production-grade deployments.

- Namespace 2 contains one pod, possibly a staging, QA, or tenant-isolated workload.

This demonstrates multi-tenancy and environment segmentation using Kubernetes namespaces. Node Pool 2 hosts Namespace 3, isolated from Node Pool 1. This pod may represent a compute-heavy batch job, a specialized service, or an environment needing different node configurations. The ability to direct workloads to different node pools based on node affinity rules, taints/tolerations, or namespace policies enables precise control over resource allocation, cost management, and workload security.

The diagram also emphasizes the decoupling of deployment orchestration from infrastructure specifics. While the DevOps pipeline automates the deployment of manifests, the AKS control plane autonomously determines optimal pod placement based on resource availability, policies, and constraints.

By structuring the AKS environment this way with CI/CD-driven deployment, centralized control plane logic, and intelligent workload scheduling across node pools and namespaces, enterprises gain a highly flexible and secure platform to deploy microservices, batch jobs, and stateful applications at scale.

Kubernetes Control Plane vs. Node Plane

A standard Kubernetes cluster consists of two distinct layers:

- The **control plane** houses critical services such as the Kubernetes API server, the controller manager, the scheduler, and the etcd data store. These components determine how workloads are deployed, balanced, and reconciled across the cluster.

- The **node plane** (or data plane) is composed of one or more **worker nodes**, virtual machines that host the running containers inside **pods**. Each node runs a kubelet agent, a container runtime (e.g., containerd), and a kube-proxy for service-level networking.

In **AKS**, the control plane is fully managed by Microsoft. It is **highly available by default**, resides in an Azure region, and is monitored and patched as part of the platform service. This removes a significant operational burden from infrastructure teams. Users are responsible only for managing **node pools**, which are sets of virtual machines running their actual workloads.

Creating a Basic AKS Cluster with Azure CLI

```
# Create resource group
az group create --name rg-gitops-demo --location eastus

# Create AKS cluster
az aks create \
  --resource-group rg-gitops-demo \
  --name my-aks-cluster \
  --node-count 2 \
  --enable-addons monitoring \
  --generate-ssh-keys

# Get credentials
az aks get-credentials --resource-group rg-gitops-demo --name my-aks-cluster
```

Kubernetes Workload Primitives

AKS is built on the same open source Kubernetes APIs as any standard cluster, meaning all fundamental constructs remain valid:

- A **Pod** is the smallest deployable unit and typically runs a single container. Multiple containers can be co-located in a single pod for tightly coupled processes, such as sidecars or log shippers.

- A **Deployment** manages the life cycle of stateless applications, automatically rolling out updates, managing scale, and ensuring the desired number of pods remain healthy.

- A **StatefulSet** handles workloads that require persistent identity and storage, such as databases and queue services.

- A **Service** defines a stable endpoint for accessing pods and abstracts load balancing within the cluster.

- An **Ingress** provides external HTTP/HTTPS access, routing requests to back-end services via path or host-based rules.

In AKS, these primitives are augmented with Azure-native capabilities. For example, a Kubernetes **Service of type LoadBalancer** will automatically provision an Azure Load Balancer with a public IP. Similarly, when an Ingress is created with annotations for Application Gateway, AKS can configure the Azure Application Gateway Ingress Controller (AGIC) to provide Layer 7 routing with WAF integration.

Node Pools and OS Options

Node pools in AKS allow workloads to be grouped and deployed based on machine characteristics. Each node pool is backed by a Virtual Machine Scale Set (VMSS), which provides autoscaling and resiliency. Architects can define multiple node pools to support heterogeneous workloads:

- A default **Linux node pool** for general-purpose applications
- A **Windows node pool** for legacy .NET Framework apps
- A **GPU-enabled node pool** for AI/ML workloads
- A **Spot instance pool** for cost-sensitive batch jobs

Each node pool can have different scaling rules, instance sizes, and zones, providing a high degree of control over **compute segmentation and optimization**.

Identity and Role-Based Access

Unlike self-hosted Kubernetes, AKS integrates tightly with **Azure Active Directory (AAD)**. This allows Kubernetes **role-based access control (RBAC)** to be mapped to **AAD users and groups**, so that cluster access is governed through the same identity provider used across the enterprise.

This model supports fine-grained control, such as

- Granting developers read-only access to the dev cluster
- Allowing DevOps engineers to deploy to staging
- Restricting access to production namespaces based on Azure AD group membership

These capabilities ensure that **identity governance in AKS aligns with corporate access policies**, reducing the risk of privilege escalation and simplifying compliance reviews.

Real-World Application: Enterprise Application Modernization

To appreciate the practical benefits of AKS, consider a global healthcare software company modernizing its patient engagement system. Originally built as a monolithic .NET application on Windows VMs, the company used AKS to migrate to a container-based microservices architecture:

- Each functional domain, appointments, messaging, and health records, was replatformed as a containerized service.

- Multiple node pools were defined: Windows nodes for legacy components, Linux nodes for .NET 8 microservices, and a GPU-enabled pool for video consultation AI services.

- Azure AD-integrated RBAC allowed the InfoSec team to lock down production namespaces, while development teams had full access to sandbox environments.

- Traffic from patients was routed through Azure Front Door to AGIC-managed Ingress controllers, ensuring low-latency and WAF-protected entry points.

This transition not only reduced infrastructure cost and deployment time but also enabled **per-service deployment pipelines**, **dynamic scaling**, and **simplified compliance auditing** using Azure-native security tools.

In summary, Azure Kubernetes Service distills the power of Kubernetes into a **cloud-native, enterprise-ready platform**. It eliminates the operational burden of managing cluster internals while enhancing governance, security, and integration across the Azure ecosystem. Understanding the architectural building blocks of AKS is foundational for leveraging its full potential in production environments.

In the next section, we shift our focus to securing AKS clusters, delving into the control plane, workload hardening, network segmentation, and runtime protections that make Kubernetes secure-by-design in Azure.

5.2 Securing Kubernetes Clusters in Azure

Security in Kubernetes must be understood as **multilayered and continuous**. It's not enough to secure ingress traffic or encrypt data at rest; security in a dynamic orchestration platform like AKS must span **infrastructure, workloads, identities, secrets, and network flows**, all while integrating with the broader enterprise posture of governance and compliance.

While Kubernetes provides the raw APIs and primitives for securing workloads, AKS enhances this foundation by embedding Azure-native controls such as Azure Active Directory (AAD), Azure Policy, Microsoft Defender for Cloud, and Azure Key Vault. Together, these services enable **Zero Trust security**, **defense-in-depth architecture**, and **compliance-aligned enforcement**.

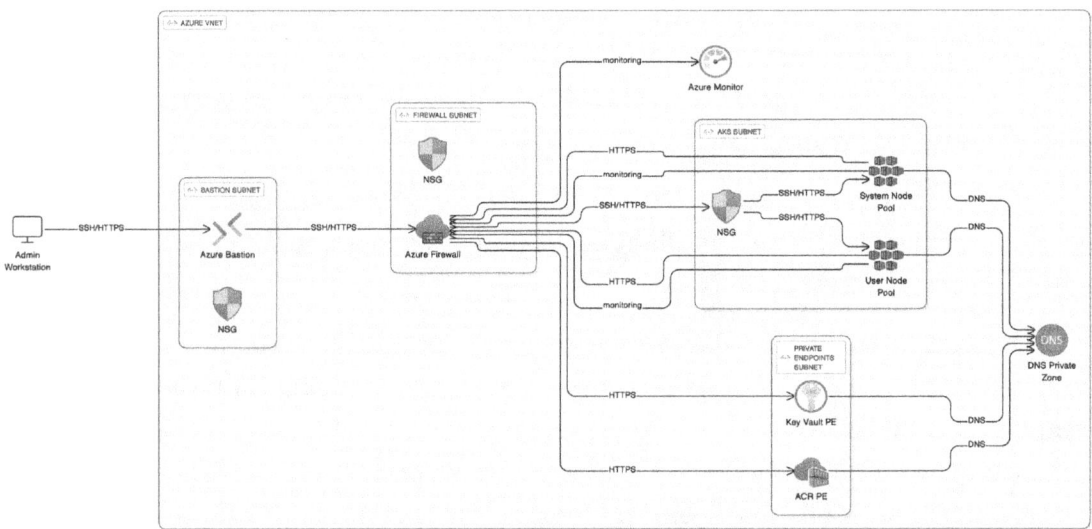

Figure 5-2. Network Security Architecture for Private AKS Cluster in Azure

This diagram illustrates a comprehensive **defense-in-depth network security design** for a **private Azure Kubernetes Service (AKS) cluster**. The architecture enforces layered security across perimeter, platform, and workload levels, ensuring secure and restricted access to all infrastructure and application components deployed within the Azure environment.

At the **perimeter layer**, inbound traffic is strictly regulated through **Azure Firewall**, which serves as the central control point for both egress and ingress flows. The firewall is deployed in a dedicated subnet within the virtual network (VNet) and configured with

explicit network and application rules. It governs all outbound connectivity from the AKS nodes and internal workloads to the internet or other Azure services, ensuring that only sanctioned traffic is allowed.

To enable **secure administrative access, Azure Bastion** is placed in the management subnet. This service provides seamless and secure RDP/SSH connectivity to the virtual machines or jump-boxes without exposing them to the public internet. All access traverses through HTTPS over port 443, ensuring encrypted and audited session management.

The AKS cluster is deployed as **private**, meaning the API server has no public endpoint and is only accessible through private network paths. Network Security Groups (NSGs) are attached to both the **AKS node pools** and **subnets** to enforce fine-grained traffic control policies at the subnet and NIC levels. NSG rules explicitly define allowed traffic patterns, such as internal cluster communication, monitoring traffic, and access from Azure Bastion or jump servers.

Access to critical services such as **Azure Key Vault** and **Azure Container Registry (ACR)** is enabled through **Private Endpoints**. These endpoints bring the respective services into the virtual network via **Private Link**, ensuring that no data traverses the public internet. DNS resolution for these services is handled via a **Private DNS Zone**, which maps the service FQDNs to their private IP addresses within the VNet, eliminating the need for public DNS queries.

All control and monitoring traffic, such as telemetry from Azure Monitor, logs to Log Analytics, and metrics from container insights, is routed through **secure diagnostic endpoints**. These may use either private endpoints or VNet integration to comply with internal traffic inspection policies.

Traffic paths are explicitly marked in the diagram:

- **SSH connections** are permitted only from Bastion to the jump-box or node VMs (if any), controlled by NSGs.

- **HTTPS traffic** flows from internal workloads to services like ACR and Key Vault through private endpoints.

- **Monitoring traffic** flows securely from the AKS cluster to Azure Monitor and Log Analytics via private or service endpoints, depending on the deployment model.

This architecture ensures that the AKS cluster and its supporting resources are **isolated, secure, and compliant** with enterprise-grade networking and security standards, adhering to the principles of **Zero Trust** and **least privilege**.

Securing the Control Plane: Public vs. Private Clusters

In AKS, Microsoft manages the Kubernetes control plane, including the API server, scheduler, and etcd database. This architecture ensures high availability and scalability out of the box, but it also requires organizations to carefully consider **how users and tools access the control plane**.

By default, AKS exposes the Kubernetes API server through a **public endpoint**, protected by Azure RBAC and Kubernetes role-based access control (RBAC). While convenient for development and CI/CD tools, this exposes the control plane to the internet, even if access is locked down to specific IP ranges.

For production environments, **private clusters** are preferred. A private AKS cluster places the API server behind a **private IP address within the virtual network (VNet)**, making it accessible only through ExpressRoute, VPN, or a jump host. Combined with Just-in-Time (JIT) access via Azure Bastion or Privileged Identity Management (PIM), this configuration limits the attack surface and conforms to **Zero Trust principles** by default.

The control plane also supports **Azure RBAC for Kubernetes Authorization**, allowing enterprises to manage access using the same identity framework applied across all Azure resources. For example, developers can be granted view-only access in dev clusters, while DevOps engineers have write access in staging, and only the release automation has rights in production.

Hardening the Node Plane and Runtime

The **node pool**, consisting of Azure VMs in a scale set, is the responsibility of the customer. This means architects must ensure these nodes are configured securely and patched regularly.

AKS supports several strategies for node hardening:

- **Automatic node OS patching and image upgrades** can be enabled to reduce exposure to kernel-level vulnerabilities.

- **Host-level threat protection** is provided via **Microsoft Defender for Servers**, which monitors suspicious behavior, privilege escalation, and known malware patterns.

- **SSH access to nodes** should be disabled unless absolutely required and, if enabled, should be protected by JIT access and restricted source IPs.

Moreover, AKS supports **node taints and tolerations**, allowing architects to schedule sensitive workloads only on trusted or isolated nodes. For example, a PCI-compliant payment processor service can be deployed to a separate node pool using hardened VM images and enhanced telemetry policies.

For runtime isolation, AKS supports both **Kata Containers** (via Confidential Computing) and **pod security context** configurations, enabling workloads to run as non-root users, drop Linux capabilities, or use read-only root file systems.

Applying a Pod Security Admission baseline policy

```
apiVersion: policy/v1
kind: PodSecurityPolicy
metadata:
  name: baseline-policy
spec:
  privileged: false
  allowPrivilegeEscalation: false
  runAsUser:
    rule: MustRunAsNonRoot
  seLinux:
    rule: RunAsAny
  fsGroup:
    rule: MustRunAs
    ranges:
      - min: 2000
        max: 2000
```

Enforcing Workload-Level Security Policies

At the workload level, Kubernetes offers extensive but often misused capabilities for access control and isolation. AKS augments this with **Azure Policy for AKS**, allowing platform teams to define governance rules declaratively.

Policies that can be enforced include

- Denying workloads without resource limits
- Blocking public container images
- Enforcing TLS for Ingress
- Restricting use of privileged containers

These policies can be scoped to namespaces, teams, or environments, and violations can be set to **audit**, **deny**, or **enforce** modes. This model is critical for large-scale platforms running **multi-team or multi-tenant workloads**, where developers must be empowered to deploy but within guardrails defined by security architects.

Additionally, **Pod Security Admission (PSA)** policies can be configured at the namespace level to prevent dangerous configurations, such as running as root, mounting host paths, or using host network interfaces.

For fine-grained admission control, **Open Policy Agent (OPA)** and **Gatekeeper** can be deployed to express complex validation logic using Rego, enabling rules like "disallow deployments using containers larger than two CPUs unless approved."

Secret Management and Identity Integration

One of the most common vulnerabilities in Kubernetes workloads is the misuse or leakage of **secret** credentials, tokens, API keys, and certificates. AKS provides several ways to secure secrets:

- **Azure Key Vault** can be used as a centralized, secure storage back end, integrated with Kubernetes via CSI (Container Storage Interface) drivers. Secrets are mounted as volumes and not persisted in etcd.
- **Workload Identity for Kubernetes** (the successor to AAD Pod Identity) allows pods to authenticate directly with Azure services using their own identity, eliminating the need to inject secrets into environment variables.

- Kubernetes native **Secrets objects** can be encrypted at rest using **customer-managed keys (CMK)** in Azure Key Vault, further enforcing data sovereignty.

In regulated industries, these capabilities enable organizations to meet strict requirements for **key rotation, access tracing**, and **least privilege** access to sensitive credentials.

Network Segmentation and Zero Trust Access

Kubernetes allows all pods to talk to each other by default. In a multi-service environment, this is unacceptable. AKS supports **network policy enforcement** using either **Azure-native policies** or **Calico**.

Architects can define

- **What service-to-service communication is allowed?**
- Whether traffic can flow across namespaces
- If certain pods must be isolated (e.g., compliance zones)

Additionally, AKS supports **Private Link** for secure, VNet-scoped access to Azure PaaS services like Cosmos DB, Key Vault, and Azure SQL. When combined with **egress firewalling** and **User-Defined Routes (UDRs)**, clusters can be locked down so that no workload has outbound internet access unless explicitly required.

Zero Trust architecture becomes possible by enforcing the principle of **default deny** and only allowing **explicit, encrypted, and authenticated** communication paths both within the cluster and with external services.

Real-World Application: Secure Healthcare AKS Cluster

A hospital network running clinical APIs and diagnostic imaging workflows on AKS must comply with **HIPAA, ISO 27001**, and local **GDPR regulations**. Their AKS security implementation includes

- A **private cluster** endpoint accessible only through VPN or Azure Bastion
- All secrets stored in Azure Key Vault, accessed via Workload Identity

- Network policies that isolate back-end services (e.g., medical record systems) from front-end portals

- Azure Policy to deny container images from public registries and enforce resource quotas

- Defender for Cloud integration to monitor threat activity and CIS benchmark compliance

- Audit logs exported to Azure Monitor and retained per compliance retention policies

The result is an AKS platform that not only runs mission-critical workloads but also stands up to external audits and internal risk assessments with confidence.

In conclusion, security in AKS is not a single configuration; it is a **framework of trust**, enforced continuously across layers. From securing access to the control plane, to isolating workloads, to protecting secrets and enforcing policy, AKS enables enterprise teams to build platforms that are secure by default, compliant by design, and resilient in the face of evolving threats.

In the next section, we will examine how AKS supports **dynamic elasticity** and **performance optimization**, ensuring that workloads can respond to changing demand without compromising cost efficiency or user experience.

5.3 Autoscaling and Performance Tuning AKS Clusters

One of the key advantages of running workloads in Kubernetes is the ability to dynamically scale based on application demand. In traditional infrastructure, this would mean manually provisioning additional servers or relying on rigid autoscale rules at the virtual machine layer. In AKS, however, scaling happens across multiple dimensions: from individual pods to the cluster's underlying compute resources, all the way to adjusting resource definitions based on usage telemetry. This **multi-tiered autoscaling** capability is what makes AKS suitable for unpredictable and high-volume enterprise workloads.

But autoscaling is not just about elasticity; it's also about **performance optimization**. Without proper tuning, you might end up with over-provisioned clusters consuming unnecessary resources or under-provisioned workloads, leading to latency spikes and failed transactions. This section explores how AKS supports intelligent autoscaling at every layer of the stack: pod, node, and resource allocation.

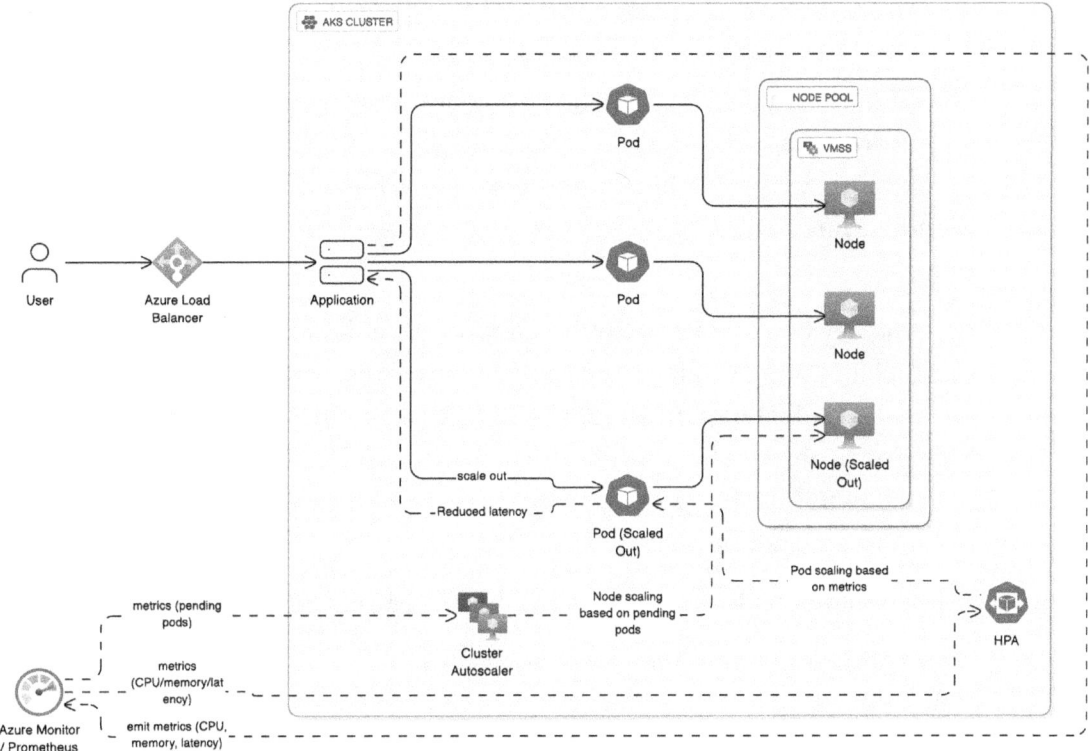

Figure 5-3. Autoscaling and Performance Tuning in Azure Kubernetes Service (AKS)

This figure illustrates a **dynamic autoscaling architecture** within a private **Azure Kubernetes Service (AKS)** cluster, showcasing how **Horizontal Pod Autoscaler (HPA)** and **Cluster Autoscaler (CA)** work together to deliver performance and cost efficiency. The architecture emphasizes a feedback-driven loop powered by **real-time monitoring** from **Azure Monitor** or **Prometheus**, enabling intelligent scaling of both application pods and underlying virtual machine infrastructure.

At the entry point, the **Azure Load Balancer** routes user traffic to the application hosted within the AKS cluster. This traffic is distributed across multiple **Kubernetes pods**, each running a replica of the application. As demand increases, pods may reach predefined resource thresholds (such as CPU, memory, or latency), triggering the **Horizontal Pod Autoscaler (HPA)** to scale out the number of pods based on those metrics.

Performance metrics originate from either **Azure Monitor** or an integrated **Prometheus** setup. These services **emit real-time metrics** such as CPU utilization, memory usage, and response latency. These metrics flow into both the **HPA** and **Cluster Autoscaler**.

- The **HPA** acts on metrics related to pod performance and scales pods **horizontally**, i.e., increasing the number of pod replicas to accommodate rising traffic. This directly helps reduce **application response latency**, as more instances become available to handle requests.

- In parallel, when additional pods cannot be scheduled due to insufficient node resources, the **Cluster Autoscaler** triggers a **scale-out of the underlying node pool**, which is backed by **Azure Virtual Machine Scale Sets (VMSS)**. The diagram clearly shows this action, with an additional **node being provisioned** inside the VMSS to accommodate pending pods.

Each component in the diagram is part of a **layered autoscaling model**:

- **User traffic** flows from the Load Balancer to the application front end.

- **Pods** scale horizontally via HPA in response to metric thresholds.

- **Nodes** scale vertically via Cluster Autoscaler based on pending pod conditions.

- **Feedback loops** connect Azure Monitor/Prometheus to the scaling controllers, completing the performance optimization circuit.

Annotated arrows trace the **metric flow**, **scaling logic**, and **performance outcomes** such as reduced latency and increased resource availability. This architecture not only ensures high availability and responsive scaling under load but also minimizes costs by deallocating resources when demand subsides.

In production-grade systems, this model supports **cloud-native elasticity**, letting the AKS cluster automatically adapt to workload demands while maintaining application health, reliability, and performance goals.

Cluster Autoscaler: Scaling Nodes Based on Pod Scheduling Pressure

The **Cluster Autoscaler (CA)** is responsible for adjusting the number of **nodes** in a node pool based on pod scheduling conditions. It watches for unschedulable pods, those that can't be placed due to a lack of CPU, memory, or GPU resources, and automatically increases the number of nodes to accommodate them.

Each **node pool** in AKS can be independently configured with

- A minimum and maximum node count
- VM size and type (e.g., compute-optimized for APIs, memory-optimized for databases)
- Scaling rules and time-based schedules (if required)

The Cluster Autoscaler also identifies **underutilized nodes** and, when safe, evicts pods to other nodes and deallocates the empty ones, saving cost without sacrificing availability.

For example, in a customer service chatbot application, pod replicas spike during the day due to user queries and drop at night. CA increases node capacity during peak periods and scales down gracefully after hours, driven entirely by workload demand.

To avoid flapping and maintain workload stability, AKS supports **scale-down delay timers** and **graceful drain configurations**, ensuring that nodes aren't terminated too aggressively.

Horizontal Pod Autoscaler (HPA): Scaling Application Instances Based on Metrics

While the Cluster Autoscaler adjusts **infrastructure capacity**, the **Horizontal Pod Autoscaler (HPA)** scales **workloads** specifically, the number of pod replicas based on real-time telemetry.

HPA supports the following metrics out of the box:

- CPU utilization
- Memory usage
- Custom metrics (e.g., queue depth, response time) via Prometheus Adapter or Azure Monitor

A web API serving dynamic content might define an HPA rule that increases the number of replicas when the average CPU exceeds 70% for 30 seconds. Similarly, a payment processing engine might scale based on queue length, increasing throughput when the order backlog grows.

AKS allows you to define **min/max replica bounds**, and HPA reconciles the desired state every few seconds. These adjustments are transparent and continuous and do not require human intervention.

Enabling Cluster Autoscaler on a Node Pool

```
# Enable autoscaling for node pool
az aks nodepool update \
  --resource-group rg-prod-platform \
  --cluster-name aks-enterprise \
  --name nodepool1 \
  --enable-cluster-autoscaler \
  --min-count 3 \
  --max-count 10
```

Defining a Horizontal Pod Autoscaler (HPA)

```
# Define Horizontal Pod Autoscaler for backend API service
apiVersion: autoscaling/v2
kind: HorizontalPodAutoscaler
metadata:
  name: backend-api-hpa
spec:
  scaleTargetRef:
    apiVersion: apps/v1
    kind: Deployment
    name: backend-api
  minReplicas: 3
  maxReplicas: 10
  metrics:
    - type: Resource
```

```
    resource:
      name: cpu
    target:
      type: Utilization
      averageUtilization: 70
```

Vertical Pod Autoscaler (VPA): Right-Sizing Container Resource Requests

In addition to scaling out with more replicas, AKS supports **Vertical Pod Autoscaler (VPA)** for optimizing the **resource allocation** of each pod. Many workloads are over-provisioned, assigning arbitrary CPU/memory values based on developer guesswork. VPA solves this by analyzing actual resource consumption and adjusting resource requests and limits accordingly.

VPA operates in three modes:

- **Off**: Provides recommendations but does not enforce changes
- **Initial**: Applies recommendations when the pod is created
- **Auto**: Continuously updates resource specs (may restart pods)

In most enterprise use cases, VPA runs in **recommendation mode** during testing and is then incorporated into CI/CD pipelines for review. This avoids runtime disruptions but still enables data-driven optimization.

For example, a predictive analytics batch job might run once nightly, but its memory needs vary with input size. VPA learns these patterns and recommends increasing memory during peak input cycles, preventing out-of-memory (OOM) errors and pod evictions.

> **Note** HPA and VPA should not target the same workload unless configured carefully, as conflicting recommendations can lead to instability. The common pattern is to use HPA for **stateless APIs** and VPA for **batch or stateful workloads**.

Node Pool Strategies and Cost-Aware Scaling

Autoscaling in AKS isn't just about performance; it's also about **cost optimization**. By using **multiple node pools**, enterprises can tailor infrastructure to match workload characteristics:

- **On-demand nodes** for critical services
- **Spot nodes** for fault-tolerant batch jobs or preemptible tasks
- **GPU pools** for inference workloads
- **Tainted pools** for security-isolated services

AKS supports configuring **Cluster Autoscaler per node pool**, which means each pool can scale independently. A media processing workload can use Spot nodes for rendering videos in parallel, accepting occasional preemptions in exchange for 70–90% cost savings.

AKS also supports **scheduled scaling** for predictable workloads, such as scaling to 20 nodes during business hours and shrinking to 5 nodes overnight, helping enterprises balance elasticity with budget control.

Application Performance Tuning Best Practices

Autoscaling alone does not guarantee performance. Applications must be designed to **scale safely and predictably**, which includes

- **Setting resource requests and limits** for every container to ensure predictable scheduling
- **Implementing robust liveness and readiness probes** to support rolling deployments and failure detection
- **Using affinity and anti-affinity rules** to co-locate or separate workloads across nodes
- **Monitoring throttling and latency** using Azure Monitor or Prometheus/Grafana

In AKS, Azure Monitor for Containers offers deep observability into CPU throttling, memory saturation, pod restarts, and disk IO performance. These insights can be fed back into autoscaler tuning and deployment strategy refinements.

Case Study: Dynamic Scaling for a Real-Time Sports App

A global sports analytics company uses AKS to run its real-time game stats engine. During major events like the FIFA World Cup, API traffic surges by over 30×. Their architecture includes

- A baseline pool of ten general-purpose nodes
- A Spot pool that can scale up to 100 nodes for real-time telemetry pipelines
- HPA scaling front-end services from 5 to 100 pods
- VPA tuning back-end services based on actual CPU/memory usage
- Azure Monitor alerts linked to Logic Apps that notify engineers if HPA or CA thresholds are breached

This architecture ensures a seamless user experience during traffic spikes while minimizing cost during off-peak hours.

In conclusion, AKS supports a comprehensive suite of autoscaling and performance tuning mechanisms that enable enterprise workloads to remain responsive, efficient, and cost-optimized under dynamic conditions. The key is to treat scaling not as a reactive fix but as a **core architectural concern**, modeled, tested, and validated like any other production-critical feature.

In the next section, we'll explore how to tie this elasticity to a repeatable deployment model using **GitOps and CI/CD pipelines** to drive safe, traceable, and auditable delivery of applications into AKS.

5.4 GitOps and CI/CD Pipelines for AKS

In traditional infrastructure, deployments are often scripted manually, with each engineer having their own environment setup and custom deployment logic. In a Kubernetes-based platform like AKS, such practices are not only inefficient; they are dangerous. The velocity of containerized development, combined with the declarative nature of Kubernetes, demands an automated, auditable, and version-controlled deployment model that integrates seamlessly into enterprise delivery pipelines.

To meet this need, AKS supports a spectrum of deployment strategies ranging from conventional CI/CD pipelines to **GitOps-based delivery models**, enabling modern DevSecOps workflows where the **source of truth is code**, the **delivery mechanism is automated**, and **environments are reproducible** on demand.

CI/CD Pipelines: Declarative Delivery Through Azure DevOps and GitHub Actions

Most enterprise teams begin their AKS journey using familiar **CI/CD pipelines** where continuous integration builds container images and continuous delivery pushes application manifests or Helm charts into the cluster.

A typical CI/CD flow for AKS includes the following stages:

1. **Build Stage**
   - Code is committed to a Git repository (Azure Repos or GitHub).
   - A CI agent triggers Docker or BuildKit to create a container image.
   - The image is pushed to **Azure Container Registry (ACR)**.
   - A software composition analysis tool scans the image for vulnerabilities.

2. **Release Stage**
   - Kubernetes manifests or Helm charts are updated with the new image tag.
   - kubectl, helm, or kustomize is used to apply changes to the target AKS cluster.
   - Azure DevOps or GitHub Actions handle approvals, rollout strategy, and artifact promotion.

This process is fully automatable and integrates with **Key Vault**, **Azure Monitor**, and **Defender for Cloud**, ensuring security, compliance, and traceability. However, it still represents a **push-based model** where CI/CD pipelines actively change the cluster state.

GitOps: Pull-Based Deployment Using Flux and Argo CD

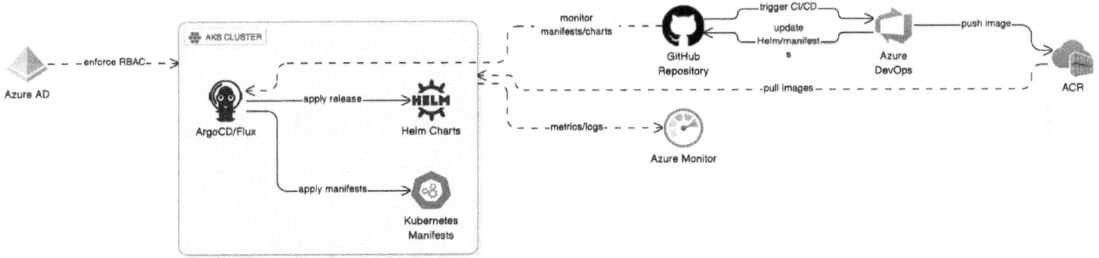

GitOps is a paradigm that reimagines application delivery as **version-controlled infrastructure reconciliation**. In this model, the Git repository is treated as the **single source of truth**, and an in-cluster GitOps agent (such as **Flux CD** or **Argo CD**) continuously monitors this repository, ensuring the cluster's state matches the desired state described in the manifests.

GitOps works in the following way:

1. Developers or platform teams push application or infrastructure YAML/Helm files to a Git repository.

2. A GitOps operator (e.g., Flux) running inside the AKS cluster watches for changes to specific directories or branches.

3. When a change is detected, the operator **pulls the change**, validates the manifest, and applies it to the cluster.

4. If the cluster diverges from the Git state (e.g., someone makes a manual change via kubectl), the GitOps agent **reverts the change,** enforcing declarative integrity.

Key advantages of GitOps include

- **Full version history** of every deployment in Git
- **Auditability** and compliance alignment through commit trails
- **Declarative drift detection and self-healing**
- Simplified **environment promotion** via pull request merges (e.g., from dev to staging to prod)

AKS offers native integration with GitOps through **AKS Flux extension** and **Azure Kubernetes Fleet Manager**, making it easy to bootstrap clusters with GitOps agents, define source-of-truth repositories, and enforce security policies.

Deployment Strategies: Canary, Blue/Green, and Progressive Delivery

Whether using CI/CD or GitOps, enterprise delivery workflows benefit from controlled rollout strategies that reduce blast radius and allow for phased validation.

AKS supports

- **Canary Deployments**: New versions are rolled out to a small subset of users or traffic. This is often implemented via Ingress controllers or service meshes like **Istio** or **Linkerd**, which control traffic weights.

- **Blue/Green Deployments**: Two environments run in parallel, blue (current) and green (new). Traffic is switched once the new deployment is verified.

- **Progressive Rollouts**: HPA or custom metrics are used to scale up new versions incrementally while watching for errors or SLO violations.

Tools like **Flagger** integrate with Flux and provide **automated progressive delivery**, halting rollouts if health metrics degrade and supporting rollback to previous versions based on service-level indicators.

For example, a banking application rolling out a new fraud detection algorithm might

- Deploy the new version to 10% of pods.

- Monitor latency and false positive rates.

- Promote to 50%, then 100% only if metrics remain stable.

- Roll back automatically if anomaly detection triggers alerts.

GitOps with Flux: Installing Flux CLI and Bootstrapping

```
# Install Flux CLI
curl -s https://fluxcd.io/install.sh | sudo bash

$env:GITHUB_TOKEN = "token"

# Bootstrap Flux
flux bootstrap github \
  --owner=your-github-username \
  --repository=nodejs-gitops-demo \
  --branch=main \
  --path=./clusters/my-aks-cluster \
  --personal \
  --token-auth
```

Managing Secrets and Environment-Specific Configurations

Secret management and environment configuration are vital in AKS. The use of plain Kubernetes Secrets is often discouraged in production due to the risk of exposure and lack of encryption by default. Instead, best practices include

- **Storing secrets in Azure Key Vault** and mounting them into pods via the **CSI Secret Store driver**
- **Injecting environment-specific configurations** using **Helm values files**, **Kustomize overlays**, or **sealed secrets**
- **Separating manifests by environment** using Git folder structures or directory conventions (e.g., apps/dev/, apps/prod/)

These approaches ensure secrets remain outside of the cluster, encrypted, access-controlled, and auditable, meeting security requirements and compliance audits with confidence.

Real-World Application: Platform Engineering at Scale

A multinational ecommerce company has standardized its internal platform on AKS using GitOps. Each microservice team manages its own Git repository with a Helm chart and values file for each environment. A central platform team maintains a base repository with

- Shared ingress configuration
- Azure Policy definitions
- Base monitoring and logging agents
- Environment-wide secrets references

The GitOps controller (Flux CD) is installed in each AKS cluster, scoped to different namespaces, and only pulls changes from approved branches. New features are deployed via pull requests, with automated policy checks, security scans, and environment promotion logic enforced via GitHub Actions and OPA Gatekeeper.

This architecture ensures full automation, team autonomy, and platform consistency while satisfying internal governance, PCI-DSS, and GDPR requirements.

In conclusion, GitOps and CI/CD pipelines are not competing paradigms but complementary approaches to **secure, traceable, and scalable delivery** in AKS. Whether you choose pipeline-based or GitOps-native deployment, the key is to treat infrastructure and application configurations as **declarative artifacts**, continuously reconciled and version-controlled, just like application code.

In the next chapter, we expand beyond deployment to focus on **observability, cost optimization, and compliance**, rounding out the operational concerns required for building and running production-grade Kubernetes platforms at enterprise scale.

5.5 Azure Container Registry Strategy for Enterprise AKS Deployments

In modern Kubernetes-based infrastructure, container registries are no longer passive image stores; they act as foundational trust anchors and life cycle orchestrators for all application workloads. For AKS clusters operating at scale with VMSS (Virtual Machine Scale Sets), adopting a disciplined approach to **Azure Container Registry (ACR)** usage, image versioning, and runtime hygiene is essential for ensuring high availability, reproducibility, and security.

This section introduces a robust registry strategy comprising **curated base images**, **immutable versioned tags**, **gRPC and crictl-driven runtime inspection**, and **automated garbage collection**, all contextualized for enterprise workloads running in AKS.

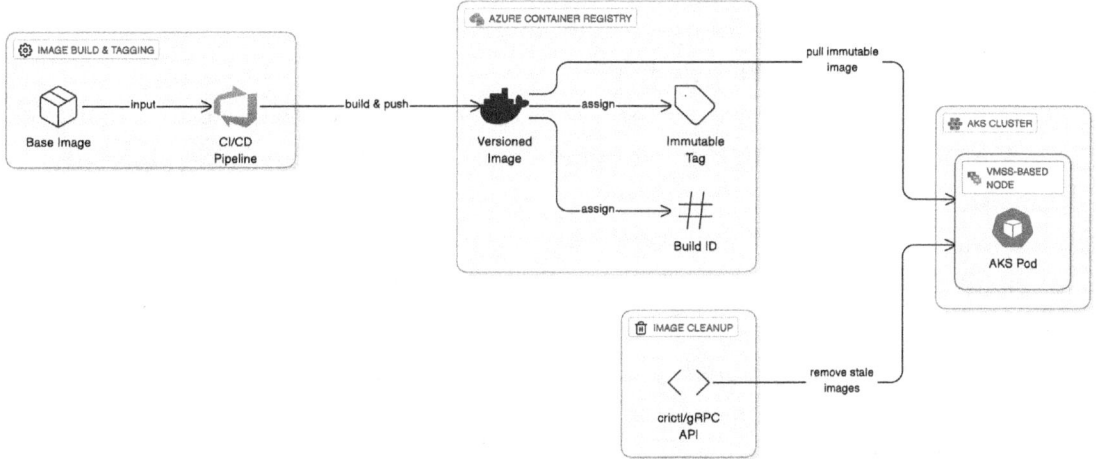

Figure 5-4. *Immutable Container Image Management and Life Cycle in AKS*

This figure represents a robust container image life cycle workflow tailored for Azure Kubernetes Service (AKS) clusters, emphasizing immutability, versioning, and secure image cleanup. The architecture enforces best practices in container build, tagging, distribution, and runtime hygiene for production-grade deployments.

The process begins in the Image Build and Tagging phase, where a base image (typically a secure, hardened OS or application runtime layer) is fed into a CI/CD pipeline such as Azure DevOps, GitHub Actions, or Jenkins. This pipeline is responsible for building the application container image, applying necessary patches, dependencies, and application code, followed by validation steps such as static analysis or vulnerability scanning.

Once the image is successfully built, the CI/CD pipeline pushes the versioned container image to an Azure Container Registry (ACR). Within ACR, each image is assigned both

- An immutable tag (e.g., v1.0.3, release-2025-06-28), which ensures that the image content never changes once published

- A Build ID (e.g., Git commit SHA or pipeline run number), providing traceability back to the source code and build context.

The AKS cluster, composed of VMSS-based nodes, subsequently pulls the immutable image from ACR. Each node then instantiates AKS Pods using the fetched image. This strategy guarantees reproducibility, security, and rollback capability, key elements in regulated or mission-critical environments.

To prevent unnecessary disk consumption and reduce attack surface area, an image cleanup process is integrated. Using the container runtime interface (CRI) via crictl or gRPC APIs, stale or unused images that are no longer referenced by active pods are automatically identified and removed from the VMSS nodes. This cleanup mechanism helps maintain node efficiency and supports ephemeral workloads in continuous deployment pipelines.

By enforcing immutable image tagging, controlled image distribution, and automated cleanup, this design adheres to core principles of DevSecOps and cloud-native security hygiene, ensuring that AKS clusters remain performant, predictable, and secure throughout the application life cycle.

Enforcing Base Image Governance Across Microservices

Enterprises typically maintain foundational base images that provide hardened environments aligned with corporate standards and compliance benchmarks. These base images act as **trusted building blocks**, including

- Preinstalled security agents and compliance scanners (e.g., Azure Defender, Trivy)

- Application runtimes like .NET 8, Node.js 20, and Python 3.11

- Custom CA certificates and telemetry agents (e.g., Azure Monitor exporters)

- Secured OS packages updated via monthly baselining

For example, a secure .NET base image may be stored as: acrcontoso.azurecr.io/baseimages/dotnet8-secure:2024.06.

CHAPTER 5   AZURE KUBERNETES SERVICE (AKS) FOR ENTERPRISE WORKLOADS

To prevent unauthorized or deprecated base images from reaching AKS, use Azure Policy for AKS to enforce allowed repositories:

```
"policyRule": {
  "if": {
    "field": "Microsoft.ContainerService/managedClusters/
    agentPoolProfiles.image",
    "notIn": [
      "acrcontoso.azurecr.io/baseimages/*"
    ]
  },
  "then": {
    "effect": "deny"
  }
}
```

This ensures all images originate from ACR-controlled pipelines and pass security validation.

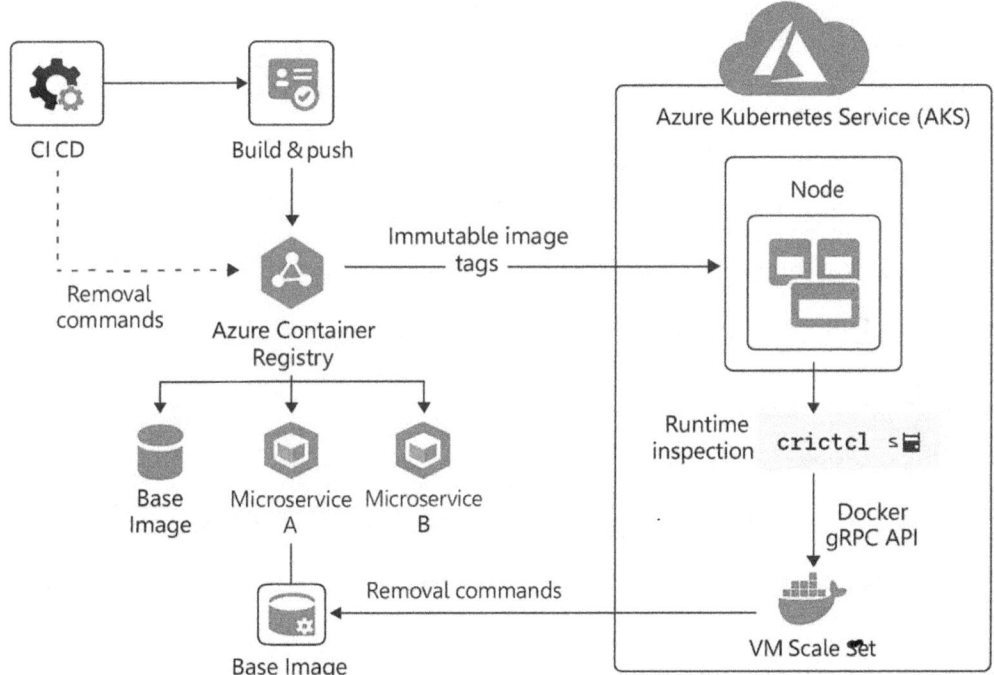

Immutable Tagging and GitOps Workflows

The cornerstone of a secure and deterministic deployment is **immutable tagging**. Each microservice image must be uniquely versioned using a combination of semantic version, build number, and commit SHA:

```
acrcontoso.azurecr.io/app-orders/backend:2.3.7-build20240527-sha4f52ac
```

Avoiding the use of mutable tags like latest prevents non-deterministic rollbacks and shields workloads from unintended image updates.

Immutable tags are published during CI/CD and should be signed or attested using tools like Notation or Sigstore for enhanced supply chain integrity. Deployment manifests (Helm or Kustomize) must pin these tags explicitly.

VMSS Runtime Hygiene with crictl and Docker gRPC

In AKS clusters using **containerd** as the runtime, Docker-specific tools like docker rm or docker image prune are ineffective. Instead, crictl becomes the authoritative interface for interacting with the runtime.

When vulnerabilities are discovered in running containers or base layers, you must ensure that

1. Affected images are not reused by the scheduler.
2. Stale image layers are purged from node disk caches.
3. Live containers are either patched or killed.

Example: Listing images on a VMSS node using crictl

```
sudo crictl images
```

Deleting a vulnerable image:

```
sudo crictl rmi acrcontoso.azurecr.io/app-orders/backend:2.3.6-sha123fa
```

You can also identify stale or orphaned containers:

```
sudo crictl ps -a | grep Exited
```

This enables administrators or automated agents (running as DaemonSets or Azure Automation agents) to perform **in-place cleanup**, reducing the risk of rehydration of vulnerable workloads from node-local cache.

For nodes still using Docker as the container runtime, the Docker gRPC API (e.g., dockerd --experimental) offers similar interfaces, but AKS is gradually phasing toward containerd-first deployments.

ACR-Driven Cleanup, Retention, and Policy Enforcement

ACR provides features to enforce repository hygiene:

- **Retention policies** to delete untagged manifests after N days
- **Repository locks** to prevent accidental deletions
- **Webhook events** to trigger downstream systems on new image pushes or deletions

A sample CLI command to clean up untagged images:

```
az acr repository delete \
  --name acrcontoso \
  --repository app-orders/backend \
  --tag none \
  --yes
```

Advanced strategies include

- Using GitHub Actions on cron triggers to identify stale image tags
- Archiving golden images to long-term Azure Blob cold storage
- Automating CVE correlation and impact analysis using Defender for Cloud

Secure Runtime Operations and Zero-Day Response

When zero-day vulnerabilities are discovered, remediation time is critical. By using a combination of

- crictl for runtime container/image management
- Azure Policy to block usage of affected tags
- GitOps pipelines for rapid rollout of patched images
- VMSS rolling upgrades (via node image or scale-set reimage)

You can achieve a highly responsive and auditable patching process.

In regulated sectors (e.g., finance, healthcare), these runtime remediations often form part of **audit trails** and security posture dashboards tracked by compliance teams.

5.6 Designing Hub-and-Spoke Network Topology for Enterprise-Grade AKS Deployments

In modern enterprise environments, network architecture serves as the backbone of secure, scalable, and governable application deployment. As Azure Kubernetes Service (AKS) becomes the standard orchestration layer for microservices and cloud-native applications, embedding it within a well-architected network design is paramount. The hub-and-spoke topology is a widely adopted architectural pattern that provides both centralized governance and isolated workload domains, making it a foundational design for production-ready AKS clusters.

CHAPTER 5   AZURE KUBERNETES SERVICE (AKS) FOR ENTERPRISE WORKLOADS

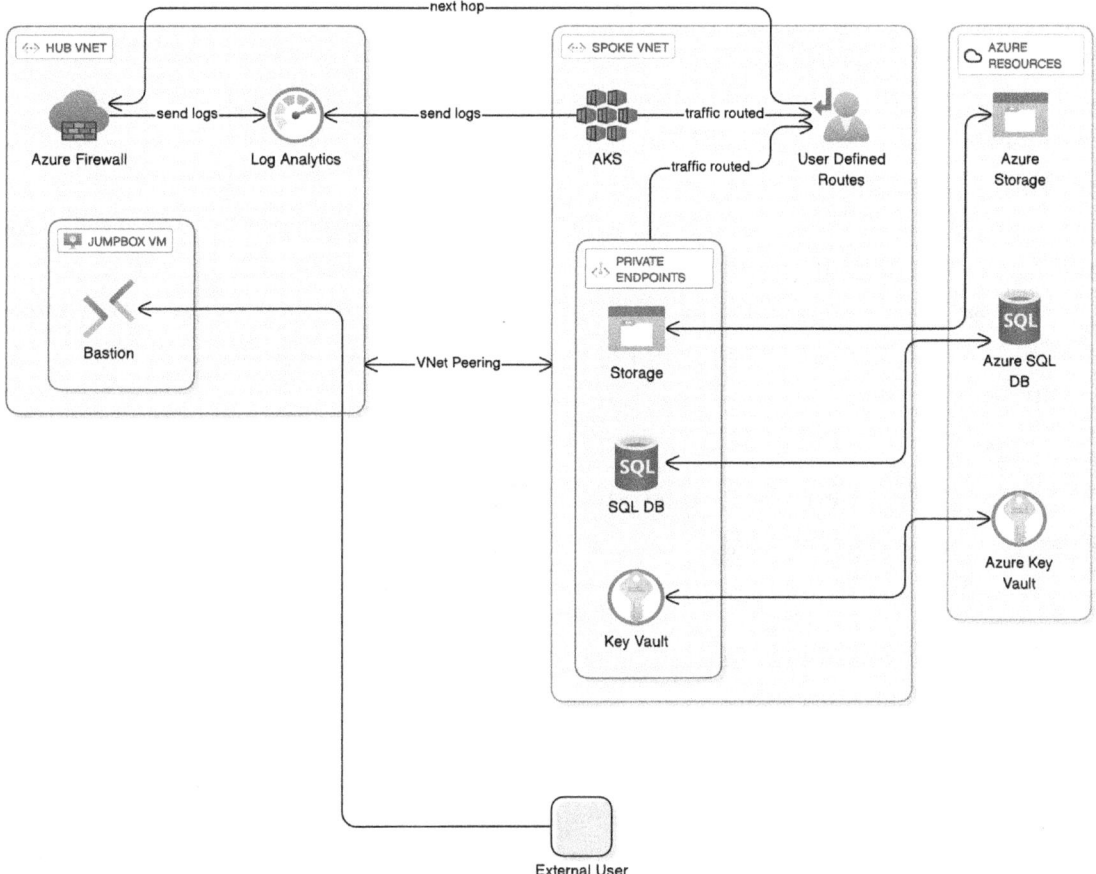

Figure 5-5. Hub-and-spoke network with Bastion, Firewall, AKS, private endpoints, and resource connectivity to Azure services

The diagram in Figure 5-5 illustrates this hub-and-spoke network architecture, where the AKS cluster is deployed into a spoke virtual network (VNet) that connects to a shared hub VNet using VNet peering. This section unpacks each component, its role in the topology, and the routing flows that ensure security, observability, and operational flexibility.

Hub Network: Centralized Control and Security Services

The hub VNet acts as the centralized control plane for cross-network services. It contains critical shared services that provide security enforcement, monitoring, and controlled administrative access. Azure Firewall is deployed in a dedicated subnet within the hub VNet. It serves as the perimeter and egress control plane, enforcing rules for both inbound and outbound traffic. It inspects and filters traffic coming from or destined to the spoke VNet. Azure Bastion provides secure SSH/RDP connectivity to internal resources without exposing public IPs. It fronts a jumpbox VM used for controlled access into private AKS nodes or other workloads. By routing administrative access through Bastion, enterprises minimize their attack surface and enforce auditability. Log Analytics Workspace is integrated for centralized telemetry ingestion. It aggregates logs and metrics from the AKS cluster, firewall, and other services, enabling holistic observability, alerting, and compliance monitoring.

All these resources reside in isolated subnets, governed by Network Security Groups (NSGs) and custom User-Defined Routes (UDRs) to ensure granular traffic control within the hub.

Spoke Network: AKS Cluster and Private Resource Integration

The spoke VNet is dedicated to hosting the AKS cluster, deployed in private mode, meaning the Kubernetes API server has no public endpoint. Within the spoke, multiple subnets support the cluster:

- A system subnet for the AKS node pools
- A pod subnet used when enabling advanced CNI networking
- A private endpoint subnet to connect to Azure PaaS services like Azure Key Vault, Azure Container Registry (ACR), and Azure Storage

Each private endpoint in this VNet connects securely to its respective Azure resource via Private Link, ensuring traffic remains on the Microsoft backbone network and is never exposed to the public internet.

The spoke-to-hub peering is configured as non-transitive, enabling the AKS nodes to reach shared services (such as Log Analytics and the firewall) without the risk of lateral traversal to other spokes. Custom UDRs within the spoke direct traffic to flow through the Azure Firewall in the hub, thus enforcing inspection and control even for intra-Azure communication.

Traffic Flow and Routing Overview

The traffic flows in this topology are directional and policy-enforced, designed to support the Zero Trust principle:

- Outbound traffic from AKS pods or nodes is routed via UDRs to Azure Firewall, which applies egress filtering policies. For example, containerized workloads accessing an external third-party API would have to conform to firewall rules.

- Ingress traffic, such as DevOps engineers connecting to a jumpbox for debugging AKS workloads, is securely tunneled through Azure Bastion, avoiding public IP exposure.

- Monitoring data from AKS nodes and workload flows into the Log Analytics workspace in the hub over private network paths, enabling visibility without compromising network boundaries.

- Private endpoints ensure that internal workloads in AKS securely access services like ACR or Key Vault without DNS leakage or unmonitored traffic.

This well-structured routing ensures that no east–west or north–south traffic is left uninspected while still maintaining performance, isolation, and flexibility.

Advantages of the Hub-and-Spoke AKS Network Design

Deploying AKS within a hub-and-spoke architecture offers numerous benefits, especially for regulated or large-scale enterprises:

> **Centralized Governance**: Firewall policies, logging, and DNS resolution can be managed centrally in the hub.
>
> **Improved Security Posture**: Bastion eliminates the need for public jumpboxes. All communication is locked within private peered networks.
>
> **Scalability**: Multiple AKS environments (e.g., dev, test, prod) can be deployed in separate spokes, all using the same centralized hub services.
>
> **Operational Efficiency**: Central Log Analytics enables unified observability, making SRE and security operations streamlined.
>
> **Network Segmentation**: NSGs and UDRs allow precise control over which traffic is allowed between nodes, workloads, and services.

5.7 Summary

In this chapter, we explored how **Azure Kubernetes Service (AKS)** serves as a foundational building block for deploying, scaling, and managing containerized workloads in the enterprise. The chapter covered the full life cycle of AKS adoption from initial cluster design to advanced operational patterns anchored in real-world requirements such as scalability, resiliency, cost optimization, and secure DevOps practices.

We began by understanding the core **AKS architecture**, including the separation of the control plane and worker nodes, and how node pools and namespaces facilitate workload isolation and resource governance. Leveraging CI/CD pipelines, DevOps teams can declaratively deploy applications into AKS using GitOps, YAML-based manifests, or Helm charts.

Next, we addressed **image life cycle management**, emphasizing the importance of using **immutable container images** stored in Azure Container Registry (ACR). We demonstrated how CI/CD pipelines build, tag, and push versioned images to ACR and how AKS pulls these images securely using private endpoints. The integration of automated image cleanup via container runtime APIs ensures disk hygiene and runtime efficiency.

The chapter then delved into **autoscaling strategies**, highlighting how **Horizontal Pod Autoscaler (HPA)** and **Cluster Autoscaler (CA)** dynamically respond to real-time performance metrics. We showed how these scaling mechanisms, powered by Azure Monitor or Prometheus, improve latency, optimize infrastructure cost, and ensure service availability during traffic surges or idle windows.

Finally, we examined a **production-grade network topology** using the **hub-and-spoke model**, where AKS clusters reside in spoke VNets and shared services like Azure Firewall, Log Analytics, and Bastion are centralized in the hub. Through VNet peering and User-Defined Routes (UDRs), this topology enforces strict traffic flows, enables centralized logging, and supports a Zero Trust security posture.

Together, these components form a cohesive blueprint for deploying AKS in enterprise environments. By following the principles outlined in this chapter, architects can ensure that their AKS implementations are **secure, scalable, observable, and cost-effective** while aligning with corporate governance policies and industry compliance standards.

CHAPTER 6

Observability, Cost Optimization, and Compliance

Building a Transparent, Efficient, and Governable Azure Infrastructure

Modern cloud infrastructure cannot be treated as a black box. The scale, elasticity, and ephemerality of services in Azure make it essential to architect systems that are **observable**, **cost-aware**, and **compliance-ready** from the ground up. In traditional on-premises systems, observability was often a second thought added reactively to debug production failures. In contrast, in the cloud-native era, observability must be designed as an intentional, first-class capability enabling platform teams to understand system behavior, optimize resources, detect threats, and demonstrate compliance before any issue arises.

This shift is not just technical; it is operational and strategic. Enterprises moving to Azure must meet aggressive uptime SLAs, manage spend across thousands of resources, and adhere to regulatory standards such as PCI-DSS, HIPAA, and GDPR. Without proper visibility into logs, metrics, alerts, and policy adherence, even the best-designed systems are vulnerable to inefficiencies, undetected failures, and compliance drift.

Azure addresses these challenges through a comprehensive ecosystem of services **Azure Monitor, Log Analytics, Application Insights, Cost Management, Azure Policy, and Microsoft Defender for Cloud**, each offering a specialized view into platform health, performance, and governance. But the true power of these tools lies in how they are integrated, automated, and governed through architecture.

CHAPTER 6   OBSERVABILITY, COST OPTIMIZATION, AND COMPLIANCE

In this chapter, we explore how to implement a **360-degree visibility model** in Azure. We will start by building a monitoring strategy using Azure-native observability tools, then apply best practices for cost optimization at scale, and finally, implement governance and policy enforcement to meet internal and external compliance needs.

6.1 Logging and Monitoring with Azure Monitor and Log Analytics

In cloud-native architectures, the ability to understand and observe systems in real time is not just a bonus; it's essential for operational reliability. Unlike traditional monolithic environments, where logs were centralized within a handful of servers, Azure infrastructure involves highly dynamic components: scaling nodes, transient containers, and platform-managed services that generate an explosion of telemetry. The challenge for modern cloud architects is to **collect, query, correlate, and act** on this telemetry to ensure visibility into performance, failures, and user impact.

Azure provides a rich, extensible observability framework centered around **Azure Monitor**, which aggregates metrics and logs from virtually every Azure service and custom application, and allows unified querying via **Log Analytics**, visualization with **workbooks**, and automated response through **alerts** and **action groups**. Together, these services empower engineers to monitor systems holistically across four core pillars of observability: **metrics, logs, traces, and alerts**.

CHAPTER 6   OBSERVABILITY, COST OPTIMIZATION, AND COMPLIANCE

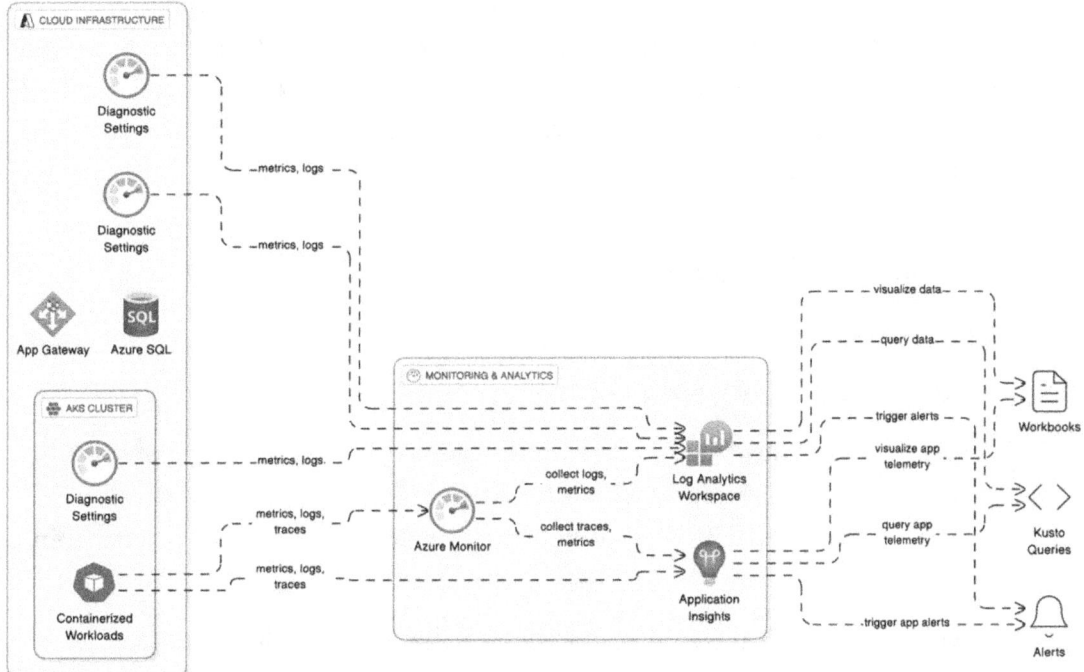

Figure 6-1. *End-to-End Observability Architecture with Azure Monitor, AKS, and Log Analytics*

Figure 6-1 presents a comprehensive view of a modern observability pipeline for Azure-based workloads, emphasizing how telemetry is collected, stored, analyzed, and acted upon in real time. This architecture focuses on Azure Monitor's deep integration with Azure Kubernetes Service (AKS), Application Insights, Log Analytics, and related services to deliver actionable insights, proactive alerts, and historical diagnostics.

At the core of the architecture is **Azure Kubernetes Service (AKS)**, hosting microservices-based container workloads. Instrumentation is enabled via the **Azure Monitor for containers extension**, which auto-injects telemetry collectors into each AKS node. These agents extract three primary signal types:

- **Metrics** (e.g., CPU, memory, node health)
- **Logs** (e.g., stdout/stderr, container logs)
- **Traces** (e.g., distributed tracing of service calls)

CHAPTER 6   OBSERVABILITY, COST OPTIMIZATION, AND COMPLIANCE

All telemetry is funneled to a central **Log Analytics Workspace**, a unified telemetry store. Diagnostic settings configured on AKS ensure both control plane logs (KubeAPI, audit logs), and data plane metrics are forwarded efficiently. **Custom queries** can be executed using **Kusto Query Language (KQL)** within the workspace.

In parallel, **Application Insights** is used for deep application-level monitoring. It captures

- **Request-response telemetry**
- **Dependency tracking**
- **Custom telemetry from app code**

This telemetry also flows into Azure Monitor and is optionally archived in the Log Analytics Workspace for correlation with infrastructure logs.

Beyond AKS and App Insights, the architecture depicts diagnostic telemetry from peripheral services like

- **Azure Application Gateway**, configured to emit access logs and performance metrics
- **Azure SQL Database**, forwarding resource health, query metrics, and error logs

These logs are streamed into the same Log Analytics Workspace via diagnostic settings, ensuring a **single pane of glass** for cross-service observability.

At the analysis and visualization layer, **Azure Monitor Workbooks** provide interactive dashboards with custom KQL queries, performance charts, and SLA tracking widgets. **Alerts** are configured using Azure Monitor's rule engine, capable of detecting threshold breaches (e.g., CPU > 80%, error rate spikes) and triggering actions like email notifications, Logic Apps, or webhook integrations.

Figure 6-1 further illustrates telemetry pathways using **arrow annotations**, clearly labeled as

- Metrics → Azure Monitor Metrics Store
- Logs → Log Analytics
- Traces → Application Insights

This visual representation enforces the idea that monitoring in Azure is not a bolt-on; it's **natively embedded across layers**, from container infrastructure to application runtime to database performance.

The result is an **observability architecture that is holistic, scalable, and cloud-native**, empowering DevOps and SRE teams to

- **Detect anomalies quickly.**
- **Correlate events across services.**
- **Investigate incidents using full-stack telemetry.**
- **Continuously improve reliability and user experience.**

Azure Monitor: The Control Plane of Observability

At the heart of Azure's observability suite is **Azure Monitor**, a service that ingests data from

- **Platform-level services** (e.g., Azure VMs, AKS, App Services)
- **Custom applications** via SDKs (e.g., Application Insights)
- **Guest OS agents** such as the Azure Monitor Agent or Log Analytics Agent
- **Diagnostic settings** from resources like Key Vault, SQL, Storage, etc.

Azure Monitor provides a **unified data pipeline**, meaning logs and metrics can be routed to **Log Analytics workspaces**, **Event Hubs**, or **Azure Storage**, depending on retention and analysis needs. This flexibility allows enterprise teams to centralize monitoring across subscriptions and regions.

For example, an AKS cluster running a .NET API back end can push container metrics (CPU, memory), Kubernetes event logs, and application traces into Azure Monitor. Platform metrics such as node disk usage or control plane availability are automatically collected from the Azure control plane, ensuring that both **infrastructure and application layers** are covered.

Log Analytics Workspaces: Queryable, Correlated Telemetry

While Azure Monitor collects data, **Log Analytics** enables teams to **interact with it intelligently**. A Log Analytics workspace is essentially a time-series database backed by **Kusto Query Language (KQL),** a powerful, SQL-like syntax optimized for querying structured and semi-structured logs at scale.

Each workspace can ingest

- **Activity logs** (control plane operations like VM create/delete)
- **Resource logs** (e.g., firewall traffic, network flows)
- **Performance metrics** from agents
- **Custom logs** from applications

A platform engineer can, for example, use KQL to identify the top ten failed container restarts in an AKS cluster in the last 24 hours:

```
ContainerLog
| where TimeGenerated > ago(24h)
| where LogEntry contains "error" or LogEntry contains "CrashLoopBackOff"
| summarize Count = count() by ContainerID, Image, LogEntry
| top 10 by Count desc
```

This ability to **search across resource types, timeframes, and dimensions** transforms telemetry from a passive log dump into a **diagnostic decision system** critical for SREs managing complex distributed systems.

Metrics, Alerts, and Visualization

Azure Monitor supports **multidimensional metrics** with numeric values that are collected at high frequency (one-minute granularity) and aggregated for real-time dashboards and alerting. Unlike logs, metrics are lightweight, continuous, and ideal for

- **SLA monitoring** (e.g., HTTP 5xx error rates)
- **Autoscaling triggers** (e.g., CPU > 70% for five minutes)
- **Health alerts** (e.g., disk queue length exceeding threshold)

For example, to monitor an Azure SQL Database's DTU usage and alert if it exceeds 85% for ten minutes:

```
az monitor metrics alert create \
  --name sql-dtu-alert \
  --resource-group rg-data \
  --scopes /subscriptions/<sub-id>/resourceGroups/rg-data/providers/
    Microsoft.Sql/servers/sql-server/databases/db \
  --condition "avg dtu_consumption_percent > 85" \
  --window-size 10m \
  --evaluation-frequency 1m \
  --action-group ag-support
```

Alerts in Azure are routed through **Action Groups**, which can notify teams via email, SMS, ITSM connectors, or trigger automation like Azure Functions or Logic Apps. This integration makes it possible to implement **self-healing systems**, where threshold breaches trigger automatic remediation such as scaling out an app service or restarting a container.

To visualize metrics, Azure offers **Workbooks**, which allow the creation of custom dashboards pulling from Log Analytics queries, metrics, and even external data. These dashboards are invaluable for NOC teams, product owners, and security auditors alike.

Application Insights: Distributed Tracing for Modern Apps

For developers building distributed microservices or serverless applications, **Application Insights** extends Azure Monitor with **request-level tracing, dependency maps, exception tracking, and usage analytics**.

- HTTP requests and exceptions are automatically captured.
- Dependencies like SQL queries, storage access, and HTTP calls are traced.
- Client-side telemetry (browser/mobile) can also be captured for end-to-end visibility.

For example, in a retail app built on Azure Functions + Cosmos DB, Application Insights will trace

1. The request to the API Gateway
2. Execution of the Azure Function
3. Call to Cosmos DB
4. Response sent to the user

If the latency increases, a trace shows **exactly which component** caused the delay: API, function, database, or network.

Application Insights can be used standalone or integrated with OpenTelemetry and popular frameworks like .NET, Node.js, and Java. Combined with **Log Analytics**, it enables a **full-stack observability story** from client to container to cloud.

Real-World Application: Centralized Monitoring for Global Enterprise

A multinational financial services firm with operations in ten regions uses Azure Monitor to centralize observability for its AKS workloads, SQL Databases, and critical services. Their setup includes

- A **central Log Analytics workspace** per region with cross-subscription telemetry aggregation

- **Workbooks** tailored for different roles: operations (SLA dashboards), security (audit trail), dev teams (performance heatmaps)

- **Azure Monitor alerts** integrated with PagerDuty for 24/7 incident response

- **Application Insights** injected into all .NET and Java microservices via CI/CD pipeline templates

- **Log sampling and retention policies** to manage ingestion cost while preserving diagnostic value

CHAPTER 6 OBSERVABILITY, COST OPTIMIZATION, AND COMPLIANCE

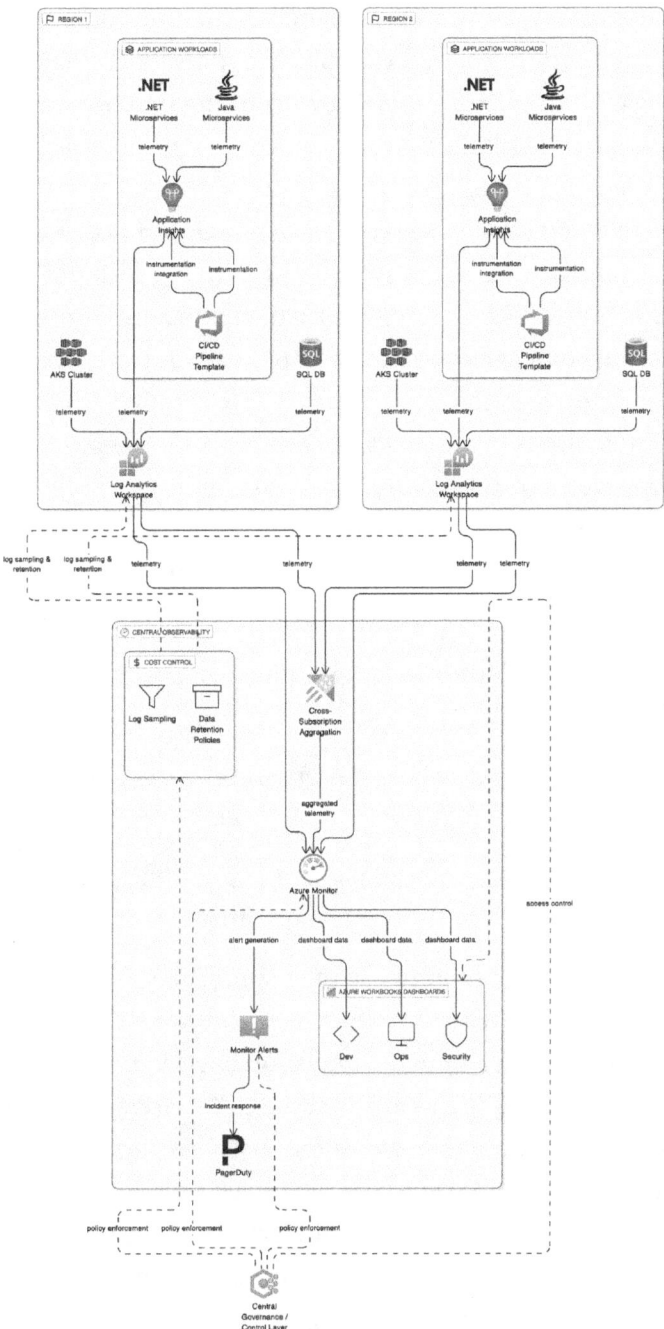

Figure 6-2. *Centralized observability with Azure Monitor, aggregating logs and metrics from multi-region microservices into Application Insights, Log Analytics, and alerting systems*

This diagram illustrates the observability architecture adopted by a multinational financial services firm operating across ten Azure regions. Each region hosts critical workloads including containerized applications running in Azure Kubernetes Service (AKS), relational data platforms such as Azure SQL Database, and distributed microservices written in .NET and Java. These workloads emit telemetry metrics, logs, and traces that are collected locally and routed to a **regional Log Analytics workspace**, ensuring low-latency ingestion and compliance with data residency requirements.

Azure Monitor acts as the central telemetry backbone, unifying data from all regional workspaces through **cross-subscription and cross-region query federation**. Workload instrumentation is automated through CI/CD pipeline templates that provision Application Insights resources and configure telemetry endpoints for each service, ensuring consistent observability across the deployment life cycle.

Ensuring full coverage without requiring manual developer intervention.

Role-based Azure Workbooks are layered on top of this data fabric, offering curated dashboards for different stakeholder personas:

- Operations teams monitor SLA compliance, resource health, and latency hotspots.

- Security teams visualize access patterns, anomaly detection, and audit trails.

- Development teams use heatmaps and dependency maps to optimize performance and troubleshoot errors.

Azure Monitor Alerts are configured with custom rules and thresholds. When triggered, they flow into **PagerDuty** for real-time incident response, enabling 24×7 monitoring across time zones. To manage ingestion costs without sacrificing diagnostic depth, the architecture employs **log sampling, data filtering, and region-specific retention policies,** ensuring only relevant telemetry is preserved long term.

The entire setup reflects a **federated yet governed observability model**, balancing local autonomy with global oversight. By enforcing standardized diagnostics configurations via policy and Infrastructure as Code, the firm ensures every new deployment, regardless of region or workload, is born observable, compliant, and production-ready.

This architecture empowers the firm's Site Reliability Engineering (SRE) teams to proactively investigate system behavior, detect anomalies early in the telemetry pipeline, and resolve potential issues before they escalate into user-facing incidents, thereby supporting sustained 99.99% service availability and satisfying regulatory logging requirements across jurisdictions.

In summary, Azure Monitor, Log Analytics, and Application Insights form the **observability backbone** of enterprise-grade Azure infrastructure. They enable deep visibility into resource behavior, rapid diagnosis of faults, intelligent alerting, and alignment with service-level objectives. When implemented correctly, they transform Azure from a deployment platform into an **introspective, self-correcting digital nervous system**.

In the next section, we will turn to cost management, where visibility meets fiscal discipline, and explore how to optimize Azure spending without compromising performance or agility.

6.2 Cost Optimization Strategies for Azure Workloads

The cloud enables unprecedented agility and scale, but it also introduces a profound shift in cost behavior. Unlike traditional data centers, where infrastructure was capital-intensive and relatively static, Azure's consumption-based model means costs can **scale up rapidly and often invisibly without architectural discipline**. For enterprises operating hundreds of services across global regions, even small inefficiencies in storage, compute, or network configurations can compound into massive financial waste.

Cost optimization in Azure is not merely about reducing spend; it is about **spending intelligently**, maximizing value while meeting SLAs, performance targets, and compliance mandates. This requires architects and operations teams to **embed financial accountability into infrastructure design**, aligning technical choices with business outcomes.

Azure offers a rich suite of native tools, **Cost Management + Billing, Advisor, Reservations, Budgets, and Pricing Calculator**, but the real power lies in how these are operationalized as part of FinOps best practices.

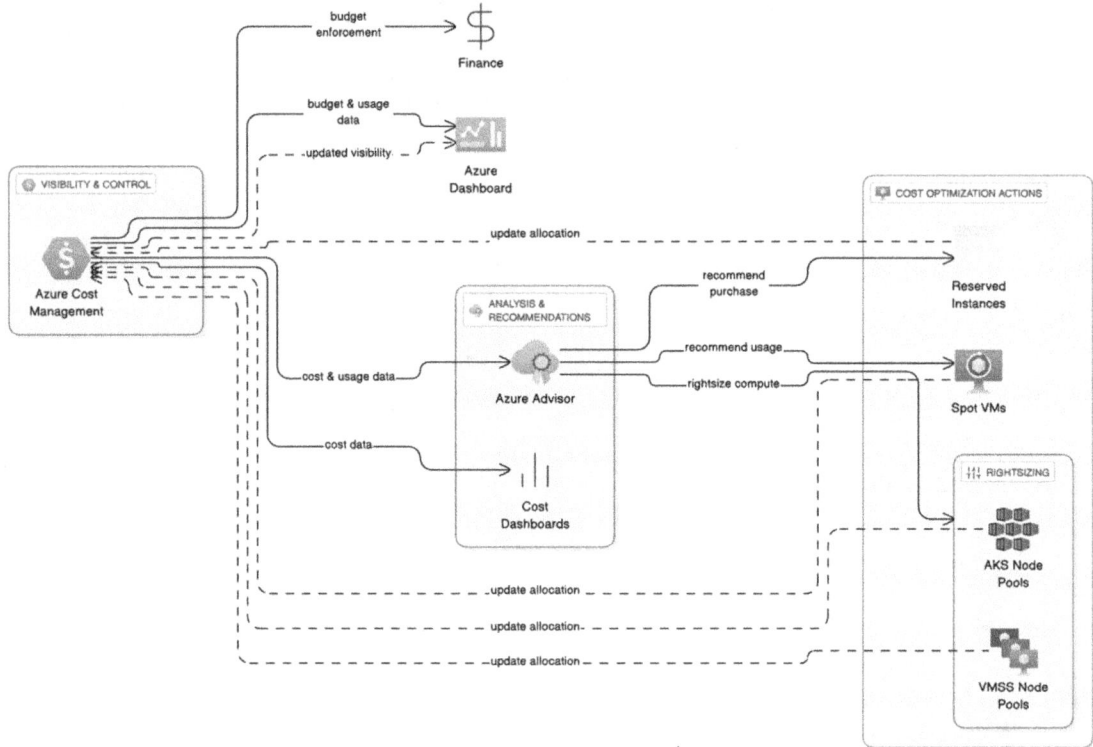

Figure 6-3. *Cost Optimization Architecture for Azure Enterprise Workloads*

Figure 6-3 visualizes a robust and multilayered architecture for enforcing cost governance and implementing proactive cost optimization in enterprise-scale Azure environments. The illustration integrates Azure-native financial tools, usage pattern analysis, and resource allocation strategies to ensure that cloud spending is both **predictable and efficient** across diverse workloads.

At the **core of the architecture** is **Azure Cost Management + Billing**, a platform service that ingests telemetry from subscriptions and resource groups across the tenant. This service provides foundational components such as

- **Cost Analysis Dashboards**: Used by engineering, finance, and leadership teams to track spend trends by service, tag, or department

- **Budget Enforcement Rules**: Allowing organizations to define monetary thresholds and trigger alerts or automation when limits are approached or breached

CHAPTER 6   OBSERVABILITY, COST OPTIMIZATION, AND COMPLIANCE

- **Forecasting Models**: Projecting future costs based on usage patterns and reserved capacity commitments

Surrounding the Cost Management hub is **Azure Advisor**, a service that continuously scans the deployment landscape for cost-saving recommendations. These include

- **Rightsizing VMs and AKS Node Pools** based on historical CPU, memory, and disk utilization

- **Eliminating idle or underutilized resources**, such as unattached disks or underused ExpressRoute circuits

- **Purchasing Reserved Instances (RIs)** for predictable, always-on compute workloads (e.g., back-end APIs or persistent databases)

- **Using Spot VMs** for interruptible, batch-style workloads like CI/CD test runners, ML model training, or video transcoding pipelines

On the **visibility and control layer**, the figure showcases

- **Finance Dashboards** shared with FinOps teams to analyze costs by cost center or project tags

- **Engineering Dashboards** with active alerts on budget limits, rightsizing opportunities, and efficiency metrics

- **Role-based access control (RBAC)**, ensuring that only authorized personnel can commit to purchases like RIs or modify production-scale VM configurations

In addition, Figure 6-3 highlights the **integration with Azure Policy**, enabling enforcement of resource SKUs, tagging strategies (e.g., costCenter, environment), and constraints on VM sizes in dev/test environments.

Together, this architecture promotes a **culture of financial accountability** where every engineering decision, be it scaling out a node pool, provisioning a VM, or selecting a database SKU, is made with real-time cost visibility and guardrails in place.

Ultimately, the cost optimization strategy illustrated in this figure is not a reactive expense-cutting measure. Rather, it is a **proactive and embedded part of the cloud operating model**, driving efficiency, sustainability, and predictability in Azure deployments across the entire enterprise.

Azure Cost Management + Billing: Visibility, Analysis, and Forecasting

The cornerstone of Azure's financial governance is **Azure Cost Management**, a suite of dashboards, reports, and APIs that help organizations monitor, analyze, and forecast cloud spend across subscriptions, resource groups, management groups, and tags.

Key capabilities include

- **Cost Analysis**: Drill-down views into cost by service, location, resource type, and tags (e.g., cost by Environment, App, or Owner).

- **Budgets**: Define monthly or quarterly thresholds and receive alerts when forecasted or actual costs exceed limits.

- **Exports**: Schedule daily or weekly cost exports to Azure Storage or Power BI for advanced reporting and cost allocation.

For example, a product team running workloads in AKS, Azure SQL, and Azure Redis can tag resources with CostCenter=RetailApps, enabling finance teams to track that unit's spend independently, even across multiple subscriptions.

```
az tag create --name CostCenter --value RetailApps

az tag update --resource-id /subscriptions/<sub-id>/resourceGroups/rg-retail/providers/Microsoft.Compute/virtualMachines/vm-retail \
  --operation merge \
  --tags CostCenter=RetailApps
```

Cost Management enables **chargeback and showback** models for internal stakeholders, making cloud economics transparent and traceable.

Azure Advisor: Proactive Cost Recommendations

Azure Advisor is an intelligent recommendation engine that analyzes resource configurations and usage patterns to identify **cost-saving opportunities**. It provides real-time guidance across five pillars: cost, security, performance, operational excellence, and availability.

Typical cost-saving recommendations include

- **Underutilized VMs**: VMs consistently running at <20% CPU can be resized or deallocated.

- **Idle App Services**: Web apps with no requests in the last 30 days can be stopped or deleted.

- **Unattached Public IPs or Disks**: Resources incurring cost without utility.

- **AKS Cluster Node Pools**: Oversized node pools can be right-sized or moved to Spot pricing.

These insights are actionable via the Azure Portal or REST APIs and can even be fed into automation tools for continuous cleanup.

Reservations and Savings Plans: Committing to Predictable Workloads

For predictable, always-on workloads (e.g., databases, middleware, shared services), Azure offers **Reserved Instances (RIs)** and **Savings Plans**, which allow customers to commit to usage over one or three years in exchange for significant discounts (up to 72% compared to pay-as-you-go).

- **Reserved Instances** are applied to VM families, regions, and sizes ideal for consistent compute usage.

- **Savings Plans** offer more flexibility, applying discounted rates across VM, App Service, Container Instances, and Functions usage for committed spend (e.g., $10,000/month for 1 year).

These can be managed at the **management group** level, ensuring enterprise-wide coverage. Azure also provides **utilization metrics** to evaluate if a reservation is being effectively consumed.

Spot VMs and Ephemeral Resources

For noncritical, fault-tolerant workloads (e.g., batch jobs, CI/CD runners, ML training), **Spot Virtual Machines** offer dramatic cost savings up to 90% by using Azure's unused compute capacity.

```
az vm create
--resource-group rg-ci
--name spot-runner
--image Canonical:0001-com-ubuntu-server-focal:20_04-lts:latest
--priority Spot
--eviction-policy Deallocate
--max-price -1
--size Standard_DS2_v2
--admin-username azureuser
--generate-ssh-keys
--location
```

While Spot VMs can be evicted at any time, they are excellent for

- Batch data processing
- Integration tests
- Caching layers
- Parallel ML model training

AKS supports Spot node pools natively, allowing specific workloads to be scheduled on these low-cost nodes using **taints and tolerations.**

Hidden Levers for Cost Efficiency

In most Azure cost breakdowns, **compute** tends to draw the most attention, but **storage and networking** often represent a **silent majority** steadily accumulating charges that go unnoticed until monthly bills reveal their weight. Yet, with a few architectural and operational best practices, these costs can be significantly curtailed without compromising performance or availability.

Tiered Storage Management

For blob storage, Azure offers **hot**, **cool**, and **archive** access tiers, each priced based on access frequency and latency expectations. Data that is infrequently accessed such as logs older than 30 days, cold backups, or regulatory archives should be transitioned from the hot tier to **cool or archive tiers**. These can reduce per-GB costs by up to **80%**, albeit with trade-offs in retrieval speed.

Life Cycle Policies

Azure Blob Storage supports **life cycle management rules** that automate transitions across tiers based on age, last access time, or custom metadata. A common pattern is to retain transactional data in hot storage for 7–30 days, then shift to cool, and eventually archive after 90 days. This ensures data remains available and compliant, while costs scale down with age.

Compression for Storage and Egress

Data compression plays a pivotal role in minimizing both storage footprint and egress costs. Whether applied during application-level data writes (e.g., GZIP for logs or CSV files) or at the content-delivery edge (e.g., Brotli or GZIP for APIs and websites), compression reduces the amount of data physically stored and transmitted. In large-scale environments, compressing telemetry, static content, or ETL outputs can yield **double-digit percentage savings** on storage and inter-region transfer bills.

Private Endpoints to Avoid NAT Egress Charges

Many enterprises mistakenly route storage or database traffic over public IPs via NAT gateways or internet-bound egress, incurring **data transfer charges**. Azure **Private Endpoints** allow traffic to stay entirely within Microsoft's backbone, bypassing public egress paths and eliminating associated costs while also enhancing security posture through network isolation.

Caching with Azure CDN or Front Door

Dynamic scaling of front-end applications often leads to excessive origin traffic. By placing static assets such as images, JavaScript bundles, or video content behind **Azure CDN** or **Azure Front Door**, you not only improve global performance through edge caching but also **dramatically reduce origin server load and outbound data costs**. Front Door also supports modern compression and intelligent caching rules, making it ideal for high-scale web applications.

> **Tip** Combine compression with caching. Serving Brotli-compressed assets via Front Door can reduce egress volumes and speed up page loads, optimizing both user experience and budget.

Storage and Network Optimization

Storage and networking costs are often overlooked, yet they account for a large portion of Azure bills. Best practices include

- **Use cool or archive tiers** for infrequently accessed blob data.
- **Enable life cycle management rules** to move files across tiers.
- **Leverage private endpoints** instead of NATed egress to reduce data transfer charges.
- **Use Azure CDN or Front Door** to cache static assets and reduce load on origin servers.

Example: Life cycle rule to move blobs to cool storage after 30 days

```
{
  "rules": [
    {
      "enabled": true,
      "name": "move-to-cool",
      "type": "Lifecycle",
      "definition": {
        "filters": {
          "blobTypes": ["blockBlob"]
        },
        "actions": {
          "baseBlob": {
            "tierToCool": {
              "daysAfterModificationGreaterThan": 30
            }
```

```
          }
        }
      }
    }
  ]
}
```

This ensures that old, rarely accessed data does not accumulate cost in hot-tier storage.

Cost Governance with Policies and Tags

Finally, true cost control comes from **preventing waste before it occurs**. Azure enables this via

- **Azure Policy**: Enforce that only specific VM sizes can be deployed or that all resources must have cost center tags.
- **Resource Locks**: Prevent accidental deletion of critical cost-driving resources.
- **Management Groups**: Apply budgets and governance across organizational units.

For example, a policy can prevent the creation of untagged resources:

```
{
  "if": {
    "field": "tags['CostCenter']",
    "exists": "false"
  },
  "then": {
    "effect": "deny"
  }
}
```

This forces developers to provide traceable metadata, enabling accurate allocation and budgeting from day one.

Cost Optimization: Multicloud Strategies and FinOps Perspective

Cost optimization isn't just about right-sizing virtual machines or removing idle resources; it also involves **strategic placement of workloads across clouds**. Enterprises adopting **multicloud strategies** often do so for regulatory, risk mitigation, or vendor lock-in reasons. However, this can have deep cost implications, both positive and negative.

From a **FinOps** standpoint, running certain workloads in another cloud provider might offer better pricing models (e.g., AWS spot instances vs. Azure low-priority VMs), but it can also introduce hidden costs in data egress, inter-cloud latency, and operational overhead. Azure Cost Management + Billing supports visibility into AWS spend using connectors, but integrating multicloud cost visibility into a single pane remains an operational challenge.

Multicloud also raises questions about **standardized observability and compliance enforcement**, which are often more seamless in a single-cloud environment. For organizations already operating in Azure, it may be more cost-effective to optimize within Azure using **Reserved Instances, Savings Plans, and Automation for shutdown schedules**, rather than spreading workloads across providers.

In summary, cost optimization in Azure is a **continuous discipline**, not a one-time exercise. It requires observability, proactive guidance, architectural decisions (e.g., reserved capacity vs. elasticity), and financial governance woven into every stage of the workload life cycle. By combining native tools with cultural alignment around FinOps, enterprises can achieve cloud cost efficiency without sacrificing performance or innovation.

In the next section, we'll examine how Azure enables **compliance, policy enforcement, and security posture management**, ensuring that cloud infrastructure is not only efficient but also **audit-ready and regulation-compliant**.

6.3 Implementing Azure Policies and Security Center

As cloud adoption accelerates across enterprises, so do concerns around **compliance drift**, **configuration sprawl**, and **security blind spots**. In traditional environments, compliance was enforced through static controls, firewalls, checklists, and manual audits. But in Azure, where infrastructure is deployed programmatically and changes occur continuously, governance must be **automated, dynamic, and proactive**.

Microsoft addresses this challenge through two tightly integrated services: **Azure Policy** and **Microsoft Defender for Cloud**. Azure Policy enables the enforcement of configuration standards at scale, while Defender for Cloud continuously assesses resource posture, detects threats, and maps environments against compliance frameworks. Together, they form the foundation of a **policy-as-code and posture-as-a-service model**, enabling organizations to embed governance into the very fabric of their infrastructure.

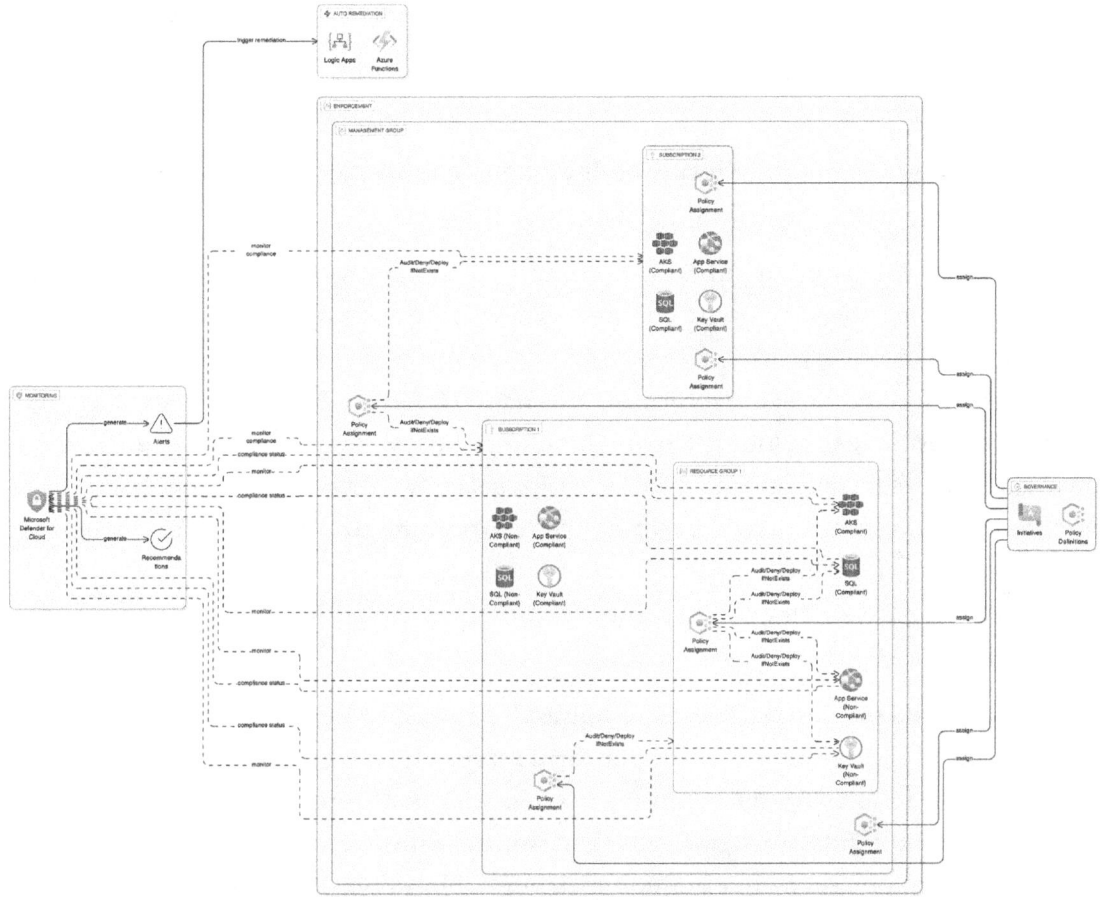

Figure 6-4. *Azure Governance Architecture with Policy Enforcement and Defender for Cloud Integration*

Figure 6-4 presents a layered governance architecture that integrates **Azure Policy** and **Microsoft Defender for Cloud** to enforce compliance, secure cloud environments, and automate remediation workflows across enterprise-scale Azure estates.

177

At the **top layer**, the architecture begins with the **Management Group hierarchy**, representing enterprise-wide governance boundaries. Azure Policy **Initiatives** (also known as policy sets) are scoped and assigned at the Management Group level. These initiatives consist of **multiple policy definitions** bundled to address thematic compliance goals such as enforcing secure networking, tagging standards, or restricting SKUs in certain environments.

Each **Policy Definition** carries a specific **policy effect**, illustrated in the diagram as

- **Audit**: To record noncompliant resources without blocking them

- **Deny**: To prevent the creation or update of noncompliant resources in real time

- **Append**: To automatically modify resource properties during deployment (e.g., adding diagnostic settings)

Policies are **inherited** by **subscriptions** and subsequently enforced across **resource groups**, enabling consistent guardrails across development, staging, and production environments.

Within a monitored **resource group**, **Microsoft Defender for Cloud** actively scans provisioned resources, including **Virtual Machines (VMs)**, **App Services**, **Databases**, and **AKS clusters**, to detect configuration drifts, security misconfigurations, and potential threats. These detections surface as **recommendations**, which are categorized by severity and compliance impact.

To transition from detection to response, the figure showcases automated remediation workflows using

- **Azure Logic Apps**: These are **manually authored by cloud administrators or central platform teams** and are triggered via **Azure Policy noncompliance events, Microsoft Defender for Cloud recommendations**, or **Event Grid notifications**. They perform predefined corrective actions such as enabling disk encryption, deploying missing diagnostic settings, or restoring baseline configurations. Azure does not create these Logic Apps automatically; they must be **developed, tested, and assigned** as remediation actions by the customer.

- **Auto-created RBAC Roles and Managed Identities**: When a **policy assignment includes a remediation task** (especially using the *DeployIfNotExists* or *Modify* effects), **Azure automatically creates a system-assigned managed identity** and a corresponding **custom RBAC role** with the least privilege necessary to carry out the remediation. This is orchestrated by the **Azure Policy engine**, ensuring that remediation operations have just enough access to succeed without granting unnecessary permissions to broader roles or requiring manual intervention.

Additionally, the diagram visualizes feedback loops:

- Policy compliance metrics are sent to **Azure Monitor**, where dashboards and alerts notify stakeholders of violations.

- Security recommendations are consolidated into **Compliance Scorecards** within the **Defender for Cloud console**, helping cloud security teams track progress against regulatory benchmarks like ISO 27001 or NIST.

The architecture is **intentionally modular**, enabling

- **Tenant-wide governance** through centralized policy definitions

- **Scoped flexibility**, allowing different subscriptions or business units to implement tailored enforcement

- **Automated feedback loops**, where policy violations immediately trigger remediation logic or security workflows

Overall, Figure 6-4 embodies a **policy-as-code and secure-by-default approach**. By leveraging Azure-native governance and security tooling in tandem, enterprises can enforce **compliance at scale**, ensure a **Zero Trust posture**, and reduce **manual overhead** associated with cloud governance, all while maintaining agility for engineering teams.

Azure Policy: Defining and Enforcing Governance at Scale

Azure Policy is a rule engine that allows organizations to **define, assign, and audit policies** across subscriptions and management groups. A policy evaluates resources upon creation or update and can enforce or audit based on defined conditions. These rules are written in JSON and can be applied across any Azure resource type.

Common use cases for Azure Policy include

- Enforcing tagging standards (e.g., CostCenter, Environment)
- Restricting VM SKUs or regions
- Requiring disk encryption or private endpoint use
- Auditing use of public IPs or HTTP traffic on web apps
- Deploying baseline resources using DINE: Use **DeployIfNotExists** to automatically create missing configurations, such as
    - Associating a default **Network Security Group (NSG)** to every newly created subnet
    - Enabling **diagnostic settings** on Azure resources to route logs to Log Analytics
    - Deploying **backup policies** or **monitoring agents** to new virtual machines

Applying **default tags** or settings when omitted by the user or pipeline

Example: Policy to deny resource creation if the CostCenter tag is missing

```
{
  "if": {
    "field": "tags['CostCenter']",
    "exists": "false"
  },
  "then": {
    "effect": "deny"
  }
}
```

This policy ensures that every resource deployed is associated with a cost center, enabling chargeback and accountability.

Policies can be grouped into **initiatives** (policy sets) and assigned at

- **Management Group Level**: Organization-wide enforcement.
- **Subscription Level**: Team or business unit boundaries.
- **Resource Group Level**: Environment-specific configurations (e.g., dev vs. prod).
- **Individual Resource Level**: While less common, Azure Policy can also be assigned directly to a **specific resource**, such as a single storage account or virtual machine. This is typically used for **high-risk or exception-driven controls** or during testing of new policy definitions in isolation before broader rollout.

Azure provides built-in policy definitions (over 400) for common scenarios, but teams can also define **custom policies** tailored to their standards.

Remediation and Compliance Tracking

Azure Policy doesn't just detect noncompliance; it also enables **remediation**. For example:

- If a storage account is deployed without encryption enabled, a remediation task can automatically enable it.
- If a resource is missing a required tag, the policy can auto-append the tag with a default value.

These capabilities make Azure Policy **stateful** and **self-healing**, reducing the need for manual correction or retroactive governance sprints.

Compliance state is surfaced through dashboards in the **Azure Policy blade**, providing

- **Compliance score by policy/initiative**
- **Trend analysis over time**
- **Remediation status and history**
- **Exemptions for specific resources**

This visibility allows security and compliance teams to respond to drift immediately while also supporting internal audits and regulatory reporting.

Microsoft Defender for Cloud: Posture Management and Threat Detection

While Azure Policy focuses on **configuration compliance**, **Microsoft Defender for Cloud** (formerly Azure Security Center) extends governance to **security posture** and **threat protection**. It provides:

- **Secure Score**: A quantified view of your current security posture, based on policy adherence, vulnerability scans, and best practice checks
- **Threat Detection**: Behavioral analytics and machine learning to identify suspicious activity, malware, brute-force attacks, and insider threats
- **Regulatory Compliance Mappings**: Automated assessments against standards such as **ISO 27001**, **PCI-DSS**, **NIST**, **SOC 2**, and **HIPAA**

Secure Score is calculated based on

- Missing or misconfigured controls (e.g., no disk encryption, public IP exposure)
- Recommendations across compute, networking, storage, and identity
- Integration with Azure Policy for enforcement

Example: Recommendations might include

- "Enable Just-in-Time access for VMs"
- "Restrict public access to storage accounts"
- "Apply system updates on VM scale sets"

Each recommendation is actionable from within the portal and can be integrated with **Azure Blueprints** or **Terraform modules** to enforce in CI/CD pipelines.

Integration with DevOps and Infrastructure As Code

Governance must be **shifted left** into the DevOps cycle. Azure Policy and Defender for Cloud support integration with

- **Azure DevOps Pipelines and GitHub Actions**: Policy evaluations during deployment
- **ARM, Bicep, and Terraform**: Policy enforcement at the IaC level using what-if and policy-as-code
- **Azure Blueprints**: Prepackaged sets of policies, role assignments, and resource templates for compliance by design

For example, a GitHub Actions pipeline can validate Bicep templates against assigned policies before deployment:

```
- name: Validate policy compliance
  run: |
    az deployment sub what-if \
      --location eastus \
      --template-file main.bicep
```

This ensures **guardrails are in place before code hits production**, reducing the risk of misconfigurations slipping through.

Real-World Application: Achieving PCI-DSS Readiness in Azure

A fintech company preparing for PCI-DSS compliance leveraged Azure Policy and Defender for Cloud to enforce

- Encryption at rest and in transit across all storage accounts and databases
- Restricted ingress using NSGs and only private endpoints for PaaS services
- Logging and diagnostic settings on every resource group
- VM vulnerability scanning using Defender agents
- RBAC policies to restrict privileged access

With Secure Score monitoring and automated remediation tasks, the team was able to achieve 97% compliance coverage across three production subscriptions, maintain continuous audit readiness, and reduce security incidents by 45% over six months.

In summary, Azure Policy and Microsoft Defender for Cloud provide a powerful governance fabric for Azure environments, ensuring configurations are compliant, security posture is continuously evaluated, and threats are detected and mitigated before they escalate. By embedding these controls into infrastructure design and DevOps workflows, organizations achieve **compliance by design**, **security by default**, and **governance as code**, the cornerstones of enterprise-grade cloud operations.

In the next section, we'll explore **compliance considerations for regulated industries**, including how Azure enables financial services, healthcare, and public sector organizations to meet legal, contractual, and industry-specific obligations.

6.4 Compliance Considerations for Regulated Industries

For many organizations, moving to the cloud is not just a matter of technological modernization; it is a strategic pivot that must be made without violating **stringent legal, regulatory, and contractual obligations**. In sectors like **financial services**, **healthcare**, **pharmaceuticals**, and **government**, compliance is not optional; it is foundational. These industries are governed by complex frameworks **PCI-DSS**, **HIPAA**, **FedRAMP**, **GDPR**, **ISO 27001**, **SOX**, and **NIST,** each with explicit requirements for security controls, data protection, auditing, traceability, and availability.

Azure is designed with these realities in mind. It provides a **compliance-ready infrastructure**, including over 100 industry certifications, and tools that allow cloud architects to **operationalize compliance controls** as code, policy, and automation. The challenge for architects is not only understanding these requirements but also designing systems that meet them dynamically, resiliently, and without impeding innovation.

CHAPTER 6 OBSERVABILITY, COST OPTIMIZATION, AND COMPLIANCE

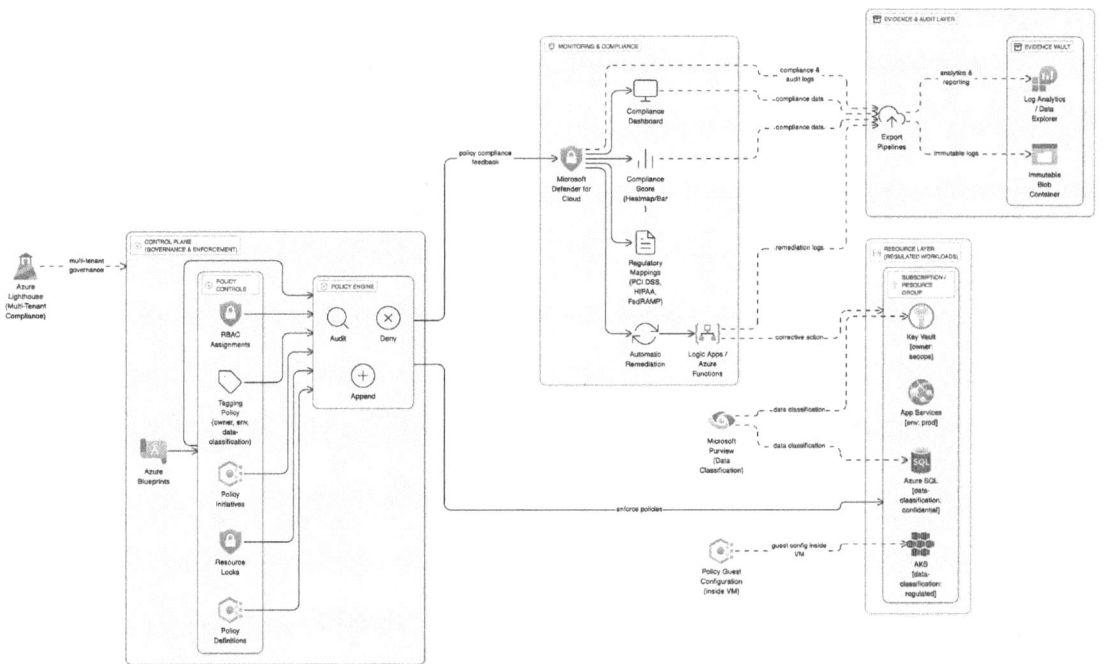

Figure 6-5. *Regulatory Compliance Architecture in Azure for Industries Like PCI DSS and HIPAA*

Figure 6-5 presents a comprehensive compliance architecture designed for **regulated industries** such as **financial services, healthcare, and government sectors** that must demonstrate conformance with frameworks like **PCI DSS**, **HIPAA**, **FedRAMP**, or **ISO 27001**. The diagram visualizes how Azure-native services collaborate to form a policy-driven, auditable, and defensible compliance posture, especially in multi-subscription enterprise environments.

At the core of this architecture is **Azure Blueprints**, a governance-centric orchestration tool that provides a **declarative framework** to package and deploy **policy assignments, role-based access control (RBAC), resource locks, and IaC templates (ARM or Bicep)** across **management groups, subscriptions, or environments** in a compliant and repeatable manner. While it shares the declarative nature of traditional **Infrastructure as Code (IaC)** tools like **Bicep** or **Terraform**, the key difference lies in **scope and intent**.

IaC tools focus primarily on defining and provisioning infrastructure resources, virtual networks, storage accounts, Kubernetes clusters, etc., often at the **resource group or subscription level**, with an emphasis on **automation, consistency, and modular reuse**.

In contrast, **Azure Blueprints** operate at a **higher governance layer**. Think of them as a **policy-backed deployment wrapper** that binds multiple governance controls, including **Azure Policy definitions, role assignments, resource locks**, and **ARM/Bicep templates**, into a single versioned artifact. This artifact can then be **assigned across multiple subscriptions**, with auditability, version control, and tracking of assignment state (published, drafted, etc.), something IaC tools do not inherently provide. Blueprints act as **compliance scaffolding**, ensuring that every new environment adheres to baseline regulatory controls from the moment it is provisioned.

Each Blueprint encapsulates

- **Pre-assigned Azure Policies** enforcing encryption-at-rest, secure networking, restricted SKUs, and diagnostic logging

- **Initiatives** mapping Azure services to regulatory controls (e.g., HIPAA §164.308 or PCI DSS 3.2.1 Req. 10)

- **Resource Locks** (ReadOnly/Delete) to protect system-critical components such as Key Vaults or Audit Logs from accidental changes

- **Mandatory Tagging Policies** (e.g., owner, data-classification, cost-center) to support audit traceability and cost attribution

Compliance Scorecards, shown in the diagram as part of **Microsoft Defender for Cloud**, continuously evaluate deployed resources against assigned regulatory standards. This includes visual heatmaps, control-by-control assessment, and a global **compliance score** that reflects posture adherence across the environment.

Complementing policy enforcement is a **security control mapping engine**, where each policy or configuration aligns with a specific compliance clause. For instance:

- Disk encryption policies map to PCI DSS Req. 3.

- VNet peering restrictions map to HIPAA safeguards.

- Log retention rules support GDPR data subject access requests.

To support **auditing and evidence collection**, the architecture integrates

- **Automated pipelines** that export **compliance telemetry** such as resource compliance states, audit logs, and activity traces into **Azure Blob Storage** or **Azure Data Explorer**.

- These artifacts are timestamped, versioned, and stored in **immutability-enabled containers** (e.g., with legal hold or time-based retention), supporting both **internal audits** and **external regulator submissions**.

In environments with high accountability, this architecture optionally integrates with

- **Microsoft Purview** for data lineage and sensitivity classification
- **Azure Policy Guest Configuration** to audit configuration drift inside VMs
- **Azure Lighthouse** for managing compliance across federated tenants (MSPs or central security teams)

Visually, the diagram organizes these layers into **Control Layers**:

1. **Preventative Controls**: Via Azure Policy and Blueprints
2. **Detective Controls**: Via Defender for Cloud and Audit Logs
3. **Corrective Controls**: Via Logic App remediations and automated enforcement
4. **Evidence Management**: Through immutable logs and export pipelines
5. **Visibility and Reporting**: Dashboards, scorecards, and Power BI integration

Logic Apps provide a serverless workflow engine that can orchestrate automated remediation actions in response to policy noncompliance events. For instance, if a storage account is detected without encryption or diagnostic settings, a Logic App can be triggered to apply the correct configuration, tag the resource, or notify the security operations team. However, it's important to note that **Azure does not automatically create these Logic Apps**. While **Azure Policy can be configured to trigger Logic App-based remediations, customers are responsible for designing, authoring, and maintaining these workflows** based on their compliance and operational needs.

Overall, Figure 6-5 demonstrates how **Azure transforms compliance from a static checkbox exercise into a living, real-time operational system**. By codifying regulatory requirements into Blueprints and Policies, continuously evaluating conformance with Defender for Cloud, and storing defensible audit evidence, enterprises can meet both the **letter and spirit of compliance** without sacrificing developer agility or operational scalability.

Azure Compliance Offerings and Certifications

Microsoft Azure has achieved compliance certifications across a wide range of international and industry-specific standards, including

- **Global**: ISO 27001, ISO 27701, SOC 1/2/3, CSA STAR
- **US Government**: FedRAMP High, DoD IL5, CJIS
- **Healthcare**: HIPAA, HITECH, HITRUST
- **Financial**: PCI-DSS, FFIEC, GLBA, SEC 17a-4
- **Europe**: GDPR, ENISA, EU Model Clauses

These certifications extend to both **platform services** (e.g., Azure SQL, AKS, Key Vault) and infrastructure services (e.g., VMs, VNets, storage). The Azure Trust Center and Microsoft Purview Compliance Manager offer access to documentation, attestation letters, and audit readiness reports.

However, certification alone is not enough. The **responsibility for compliance is shared.** Microsoft handles the platform (physical security, hypervisor, host OS), while customers must ensure secure application design, data classification, encryption, access control, and retention policies.

Key Compliance Design Considerations

To operate in a regulated environment, cloud architects must design Azure workloads that adhere to specific technical and operational requirements. These include

1. **Data Residency and Sovereignty**

    Workloads may need to ensure data does not leave a specific geography (e.g., EU for GDPR, India for IRDAI).

    - Use **region-specific services** (e.g., Azure Germany, Azure China, Azure Government).
    - Store data in **geo-restricted storage accounts**, using region-pinned replication (e.g., GRS within EU only).

CHAPTER 6　OBSERVABILITY, COST OPTIMIZATION, AND COMPLIANCE

2. **Encryption at Rest and in Transit**

    Virtually all compliance frameworks require encryption of sensitive data.

    - Use **Azure Storage Service Encryption (SSE)** with **customer-managed keys (CMK)** via Azure Key Vault.

    - Enforce **TLS 1.2 or higher** for all network communications using Application Gateway or Azure Front Door.

    - For databases, enable **Transparent Data Encryption (TDE)** with bring-your-own-key (BYOK) configurations.

3. **Access Control and Audit Trails**

    Least privilege and traceability are critical for internal controls and audits.

    - Implement **Azure AD PIM** to provide Just-in-Time access to production.

    - Use **role-based access control (RBAC)** with scoped assignments (e.g., resource group level).

    - Enable **Azure Activity Logs**, **Diagnostic Logs**, and **Log Analytics** for immutable, queryable trails.

4. **Retention, Archival, and Deletion Policies**

    Regulations like SOX and GDPR define strict rules for data retention and erasure.

    - Configure **Storage Life Cycle Management Policies** for long-term retention in the Cool or Archive tier.

    - Define **time-based retention (TBR)** and **legal hold policies** for Azure Blob storage.

    - Use **Azure Purview** and **Microsoft Purview Data Life Cycle** to manage sensitive data classification and automated deletion workflows.

5. **Business Continuity and High Availability**

   PCI-DSS and HIPAA mandate recoverability and uptime for critical services.

   - Design multi-region disaster recovery using **Azure Site Recovery (ASR)**.

   - Configure **Availability Zones** for regional fault tolerance.

   - Test DR runbooks regularly and ensure **Recovery Time Objective (RTO)** and **Recovery Point Objective (RPO)** meet contractual expectations.

Compliance Tooling in Azure

Azure provides several first-class tools to enable compliance visibility and enforcement:

- **Microsoft Defender for Cloud: Regulatory Compliance Blade** – Maps current posture to frameworks like PCI-DSS, ISO 27001, and NIST.

- **Azure Policy: Built-In Initiatives for Compliance** – Dozens of policy sets tailored to specific frameworks can be assigned to subscriptions or management groups.

- **Azure Blueprints**: Combine policies, role assignments, ARM templates, and resources into reusable packages. For example, a PCI-DSS blueprint might enforce

   - Logging on to all resources

   - RBAC instead of shared credentials

   - Network isolation for payment systems

These blueprints allow platform teams to **standardize secure environments** that comply with internal audit controls and external legal mandates without manual intervention.

Compliance: Leveraging DINE Policies for Proactive Governance

Compliance enforcement in Azure has evolved beyond static audits and manual reviews. Today, **Azure Policy** enables real-time, programmatic enforcement of organizational standards across subscriptions, regions, and environments. One of the most powerful and often underutilized constructs is the **DINE (DeployIfNotExists)** policy effect.

A DINE policy allows organizations to **automatically deploy a required resource** such as diagnostic settings, log retention rules, encryption, or Network Security Groups if it doesn't exist. This is especially critical for platform teams managing central controls across multiple application teams and environments.

For example, a DINE policy can ensure that every storage account has diagnostics enabled and logs flowing to a central Log Analytics workspace. If an application team forgets to configure this, the DINE policy intervenes at deployment time and enforces compliance without blocking the deployment, thus preserving developer agility while ensuring organizational guardrails are respected.

DINE policies are often used for

- Enabling diagnostics and auditing
- Enforcing tag governance
- Ensuring backup configurations
- Automatically associating NSGs to subnets

These policies represent a shift from **detect-and-respond** to **auto-remediate-at-deploy**, aligning perfectly with the **"compliance-as-code"** philosophy.

> **Tip** Incorporate DINE policies as part of your **Landing Zone architecture** and apply them at the **Management Group** level to ensure compliance baselines are enforced across all child subscriptions.

CHAPTER 6   OBSERVABILITY, COST OPTIMIZATION, AND COMPLIANCE

Real-World Application: Compliance-First Azure Platform for Healthcare

A digital health startup serving hospitals across Europe designed its platform on Azure with GDPR and HIPAA compliance as core architectural drivers. Their approach included

- **Data residency controls** using region-specific storage accounts in West Europe and Germany North

- **Encryption enforcement** using customer-managed keys in Azure Key Vault integrated with all storage and SQL services

- **Audit trails** for every administrative action, retained for seven years in Log Analytics and exported to secure long-term blob storage

- **Secure development pipelines** with Azure DevOps using Azure Policy to deny noncompliant resources during deployment

By using Azure's compliance blueprints, secure score recommendations, and DevOps integration, the startup passed its external GDPR audit in its first attempt and achieved ISO 27001 certification within nine months of go-live.

6.5 Summary

In a world where cloud infrastructure is the foundation of modern digital enterprises, observability, cost control, and compliance are not peripheral concerns; they are architectural imperatives. Chapter 6 has established a blueprint for embedding these critical pillars directly into the fabric of Azure environments.

Operational excellence begins with end-to-end observability, not as an afterthought but as a first-class design concern. Azure Monitor, Log Analytics, and Application Insights collectively offer a telemetry backbone capable of instrumenting everything from microservices and databases to platform resources and security controls. We explored how enterprise-scale organizations structure their monitoring layers, aggregating metrics, logs, and traces across subscriptions and regions while tailoring workbooks and alerts for distinct personas, from SREs to CISOs.

Cost optimization is not simply a matter of post hoc budgeting. With Azure Cost Management APIs, budgets, anomaly detection, and tagging enforcement policies, we

CHAPTER 6 OBSERVABILITY, COST OPTIMIZATION, AND COMPLIANCE

saw how proactive visibility into spend can be woven into DevOps workflows. Intelligent autoscaling, rightsizing recommendations, and reserved instance planning form the basis of cloud financial operations (FinOps) maturity, enabling organizations to balance performance with fiscal responsibility.

Compliance, however, is where architecture meets accountability. We examined how regulated industries use Azure Policy, Blueprints, and Defender for Cloud to enforce continuous compliance, embedding security controls, governance policies, and evidence collection pipelines into their deployment pipelines. Rather than retrofitting controls, we emphasized the importance of engineering trust into the platform from the ground up. Immutable storage, resource locks, secure baselines, and audit-ready dashboards ensure that the system not only runs but proves its integrity at every layer.

In summary, Chapter 6 affirms that operational resilience, financial efficiency, and regulatory adherence are not opposing forces; they are outcomes of deliberate, well-instrumented architecture. When observability, cost management, and compliance are treated as code declarative, repeatable, and auditable, enterprises can innovate confidently, scale securely, and pass scrutiny without friction.

As we move into Chapter 7, we transition from architectural principles to practical proof points. The next chapter presents real-world case studies that showcase how global enterprises have implemented the IaC patterns, governance strategies, and cloud-native services covered so far, demonstrating the impact of disciplined Azure architecture on performance, compliance, and business outcomes.

CHAPTER 7

Real-World Case Studies and Future Trends

Architecting Azure Infrastructure with Evidence, Experience, and Emerging Insight

Cloud transformation is no longer a question of possibility, but one of velocity and assurance. As enterprises race to modernize their technology foundations, the conversation has shifted from *"Should we move to the cloud?"* to *"How quickly can we migrate, and how securely can we operate?"* By 2025, global cloud investment is forecasted to surpass the $1 trillion mark, with Microsoft Azure emerging as a dominant force thanks to its comprehensive suite of enterprise-grade services, deep hybrid integration, and rigorous compliance credentials.

However, for large and complex organizations, embracing Azure goes far beyond the mere act of provisioning resources. True cloud adoption requires a holistic transformation strategy, one that aligns digital infrastructure with measurable business objectives, addresses regional and industry-specific regulatory mandates, streamlines operational governance across global teams, and fosters agility without compromising control. In this landscape, Azure is not just a cloud provider; it becomes a platform for reshaping the very fabric of how modern enterprises operate, innovate, and compete.

In this chapter, we journey beyond principles and practices into the dynamic realm of real-world Azure deployments. Through in-depth case studies, we uncover how leading enterprises in sectors such as finance, healthcare, and global retail have architected, secured, scaled, and evolved their Azure environments to meet the demands of their digital agendas. Each case offers not just architectural patterns but also organizational strategies, governance models, DevOps pipelines, and lessons learned the hard way.

This chapter also peers into the future: exploring emerging patterns in AI-ready infrastructure, confidential computing, sustainability in cloud architecture, and the evolving role of edge computing within Azure's global mesh. These trends, while still crystallizing, are already shaping architectural decision-making across forward-looking organizations.

We begin with a foundational question every architect must face in enterprise transformation: What does it truly mean to migrate "at scale" without losing control, performance, or security?

7.1 Case Study 1: Large-Scale Enterprise Migration to Azure Financial Services at Scale

In the high-stakes world of financial services, every millisecond counts. Latency impacts trade execution. Uptime affects customer trust. Security lapses trigger regulatory scrutiny and brand erosion. When a global investment bank with over $800B in assets under management decided to move its core workloads to Azure, the goals were ambitious but precise: reduce infrastructure operational costs by 40%, accelerate developer onboarding and environment provisioning, and meet region-specific data sovereignty requirements across North America, Europe, and Asia-Pacific. This case study dissects how this financial institution rearchitected its core infrastructure on Azure using landing zones, Zero Trust principles, and federated governance while maintaining strict compliance with global financial regulations such as GDPR, PCI DSS, and MAS TRM.

Business Drivers and Transformation Goals

The bank's data centers, spread across five continents, were nearing the end of lease cycles. Operational inefficiencies and legacy licensing costs mounted, while internal developer teams faced long lead times often two to three weeks for basic test environment provisioning. Additionally, regulators in Singapore, Germany, and Canada were tightening requirements around in-country data processing and resilience planning.

Azure was selected not merely for compute elasticity but for its unique advantages in

- Multi-region availability and compliance certifications
- Native integration with Microsoft's security suite (Defender, Sentinel)

- First-class support for hybrid identity and on-premises connectivity
- Scalable PaaS offerings for event-driven financial workflows and secure APIs

The transformation goal was to enable **regionally compliant, globally consistent, secure-by-design landing zones** for each line of business (LOB): wealth management, retail banking, commercial lending, and capital markets, each with its own budget, policy, and RBAC boundary.

Designing Scalable Landing Zones

Before any workload migration began, the Cloud Center of Excellence (CCoE) implemented Azure Landing Zones using the Microsoft Cloud Adoption Framework (CAF) and Terraform automation pipelines. The architecture enforced strong separation between platform services and application environments, aligned with enterprise policy control.

Key features included

- **Management Group Hierarchy**: Root ➤ Corp (non-prod) / Prod ➤ LOB-specific management groups (e.g., RetailBanking, TradingDesk).

- **Policy Assignments**: Applied at the Prod level using Azure Policy initiatives. Deny rules for public IPs, mandate tag compliance, and automated backup policies.

- **Subscription Provisioning**: Each LOB received multiple subscriptions: Platform, Identity, Network, and Apps, following the guidance of the Enterprise-Scale reference architecture.

- **Role-Based Access Control (RBAC)**: Defined at the management group and resource group levels. LOB architects could manage VNets but could not modify the shared hub-spoke backbone.

CHAPTER 7 REAL-WORLD CASE STUDIES AND FUTURE TRENDS

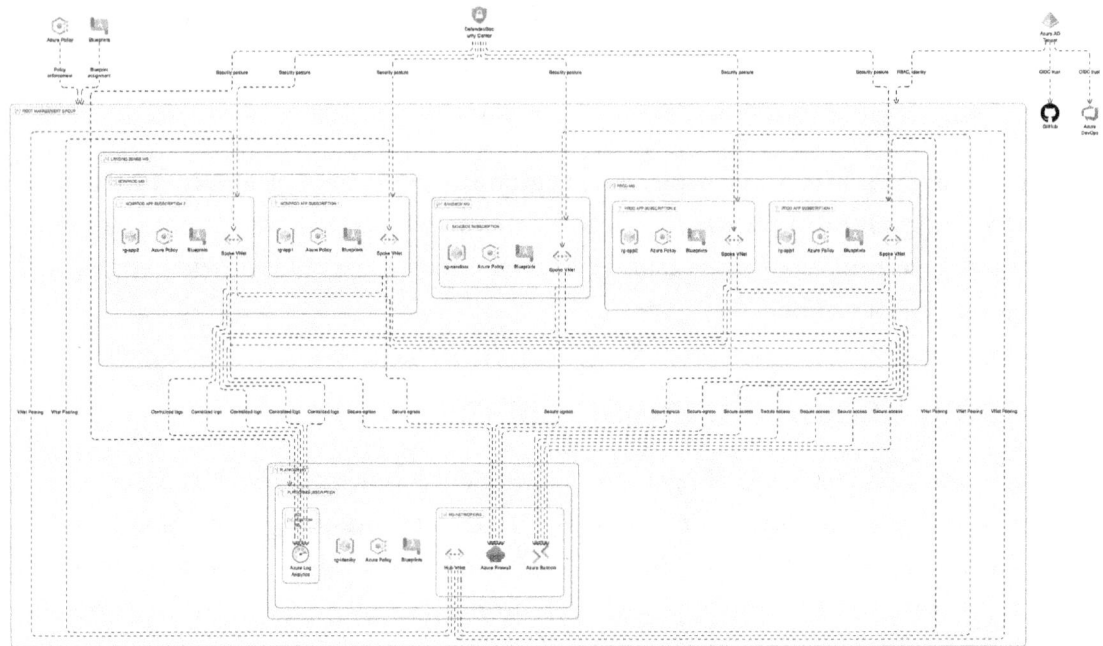

Figure 7-1. *Top-Level Landing Zone Structure*

Figure 7-1 visualizes the top-level structure of an Azure Landing Zone, an enterprise-grade foundation that codifies cloud governance, network topology, security baselines, and environment boundaries. It forms the bedrock of a scalable, compliant, and production-ready Azure footprint for large organizations undergoing digital transformation.

At the highest tier, the architecture is segmented by **management groups**, aligned with organizational boundaries such as **Platform, Corp, Online, DevOps,** and **Security**. These management groups enforce **role-based access control (RBAC), Azure Policy inheritance**, and **budgetary scopes** across multiple subscriptions and business units. The top-level management group (often named contoso-root or tenant-root) applies universal policies such as resource naming standards, region restrictions, or allowed resource types, ensuring consistent guardrails at scale.

Beneath the **management group** hierarchy, the Azure enterprise environment is typically segmented into a set of purpose-driven **subscriptions**, each aligned to a specific domain of responsibility and operational scope. This deliberate separation enhances governance, simplifies cost tracking, and enforces the principle of least privilege across different functional areas of the cloud platform.

The **Connectivity Subscription** serves as the backbone of the organization's network architecture. It hosts foundational components such as **hub virtual networks**, **Azure Firewall**, **VPN Gateways**, and **ExpressRoute circuits**. These services establish secure connectivity between on-premises datacenters, Azure regions, and workload-specific spokes, embodying the widely adopted **hub-and-spoke** topology for scalable and centralized network management.

The **Identity Subscription** is responsible for managing the infrastructure that supports identity and access operations. This includes the deployment of **Azure AD Domain Services**, **Jumpboxes**, or **Bastion Hosts** for secure administrator access and the configuration of **Privileged Identity Management (PIM)** to enforce just-in-time elevation and auditability of privileged roles. It's important to understand that **Azure Active Directory (Azure AD)** itself is a **tenant-scoped service**, not confined to any single subscription. The Identity Subscription, therefore, complements Azure AD by hosting supporting services that enforce access governance across all subscriptions under the same tenant.

The **Management Subscription** centralizes shared monitoring and automation tooling. This includes **Log Analytics workspaces**, **Azure Monitor agents**, **Update Management**, and **Azure Automation accounts**. By consolidating these services in one location, the platform team can ensure uniform telemetry collection, alerting, and compliance automation across the enterprise estate.

Finally, **Landing Zone Subscriptions**, typically segmented into environments such as **Development, QA, and Production**, are provisioned for application workloads. These subscriptions are governed by strict policies, including **resource locks**, **role-based access controls**, and **restricted resource provider registration**. They operate within clearly defined boundaries to prevent drift and enforce environment-specific controls while consuming shared platform capabilities from the Connectivity, Identity, and Management subscriptions. Each landing zone is deployed via **Infrastructure as Code (IaC)** templates, typically using Bicep or Terraform, managed through a centralized DevOps pipeline. These templates configure

- **Virtual networks (VNets)** with Network Security Groups (NSGs)
- **Route tables and UDRs** to control traffic flow between zones
- **Azure Key Vaults** for secrets management
- **Log Analytics and Application Insights** for observability
- **Azure Policy Assignments** to ensure compliance (e.g., encryption at rest, tagging, location constraints)

CHAPTER 7 REAL-WORLD CASE STUDIES AND FUTURE TRENDS

Figure 7-1 also outlines the integration of **platform services** that span across landing zones. This includes

- **Azure Blueprint definitions**, which deploy preapproved resource configurations and policies in a consistent manner

- **Defender for Cloud**, configured centrally but reporting across all subscriptions to detect threats and misconfigurations

- **Sentinel Workspaces** for Security Information and Event Management (SIEM), enabling proactive security analytics and incident response

The architecture emphasizes **modularity and decoupling**: platform services are decoupled from application workloads, enabling application teams to operate autonomously within their subscriptions while adhering to centralized controls. However, this centralization introduces a **critical risk surface**: if the **platform subscription is compromised**, the blast radius could extend to all dependent application environments. For instance, if the **Identity Subscription** hosting Bastion Hosts, PIM, or Azure AD DS is breached, attackers may gain indirect access paths to downstream services. Similarly, compromise of the **Connectivity Subscription**, which contains core networking components like Azure Firewall or Private DNS Zones, could enable traffic interception or denial of service across multiple workloads.

In addition to the technical layout, **Figure 7-1** illustrates **life cycle environments** Dev, QA, Staging, and Production mapped to different landing zones, with automated promotion and policy-driven controls enforced at each stage. **Network isolation**, **service endpoint controls**, and **Private Link** configurations ensure that sensitive data remains within the organization's trusted boundaries.

The result is a secure-by-default, compliant-by-design Azure environment that can be scaled horizontally by onboarding new business units or workloads without rearchitecting the core foundation. This aligns with Microsoft's **Cloud Adoption Framework (CAF)** and **Enterprise-Scale Landing Zone (ESLZ)** methodology.

By visualizing this structure in **Figure 7-1**, the reader gains a clear understanding of how governance, operations, and workload deployment are harmonized in a robust Azure environment designed for growth, agility, and control.

Implementing the Hub-and-Spoke Network Model

To satisfy both security segmentation and compliance mandates, a **hub-and-spoke virtual network architecture** was established in each primary region (East US, West Europe, Southeast Asia). The shared hub hosted

- Azure Firewall Premium with TLS inspection and threat intelligence
- Azure Bastion for secure VM access without public IPs
- Private DNS Zones and custom domain forwarding
- VPN Gateways and ExpressRoute circuits connected to the bank's MPLS backbone

Each LOB app was deployed into a dedicated spoke, with VNet peering back to the hub. Traffic inspection occurred centrally, and NSGs on the spoke enforced microsegmentation down to the subnet level.

Terraform Code Snippet: Spoke VNet Peering Configuration

```
resource "azurerm_virtual_network_peering" "spoke_to_hub" {
  name                      = "spoke-to-hub-peer"
  resource_group_name       = azurerm_resource_group.spoke.name
  virtual_network_name      = azurerm_virtual_network.spoke.name
  remote_virtual_network_id = azurerm_virtual_network.hub.id
  allow_forwarded_traffic   = true
  allow_gateway_transit     = false
  use_remote_gateways       = true
}
```

This peering enables spoke VNets to access shared services in the hub while maintaining clear L3 isolation from other spokes.

Identity and Access Management

Azure Active Directory (Azure AD) was federated with the bank's on-premises AD using Azure AD Connect. MFA policies were enforced using Conditional Access, and PIM (Privileged Identity Management) was mandated for all role elevation actions.

CHAPTER 7 REAL-WORLD CASE STUDIES AND FUTURE TRENDS

Workload identity for app services, container workloads, and Azure Functions leveraged **Managed Identities**, avoiding the need for secrets in pipelines or code.

RBAC inheritance and Just-in-Time (JIT) access were defined centrally through Azure Blueprints and Azure Lighthouse for shared visibility across global regions.

CI/CD and Infrastructure as Code

All platform resources were provisioned via Terraform. Application teams used Azure DevOps with environment-specific pipelines that referenced centralized modules. Terraform state was stored in encrypted Azure Storage with SAS access.

Deployment stages included

- Plan and validate against azurerm_policy_assignment.
- Apply only after approval from Platform Ops.
- Post-deployment security validation with AzSK and PSRule.

Each code push triggered release gates that integrated with Microsoft Defender for DevOps to ensure no high-severity vulnerabilities were introduced into the release pipeline.

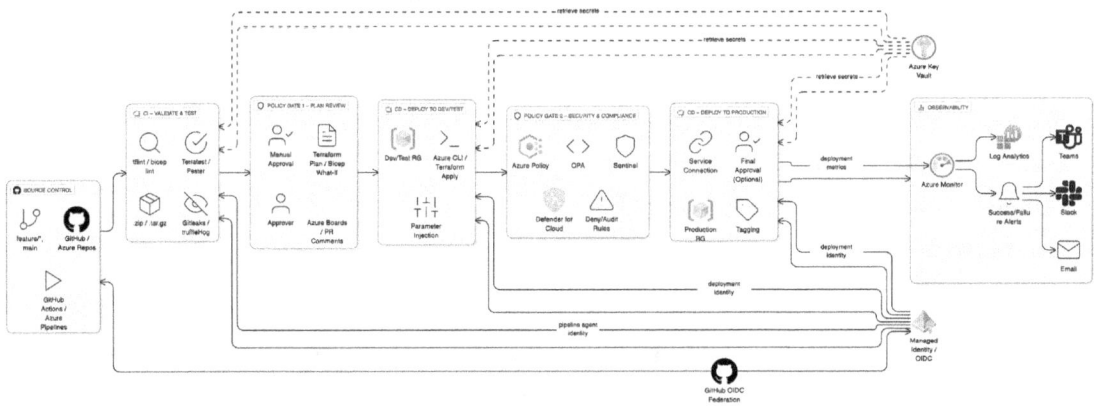

Figure 7-2. *DevOps Pipeline and Integration with Policy Gates*

Figure 7-2 presents a comprehensive view of a modern Azure DevOps CI/CD pipeline, enriched with integrated policy gates, compliance checks, and multistage approval workflows tailored for secure enterprise infrastructure delivery. This diagram serves as the blueprint for orchestrating Infrastructure as Code (IaC) deployments using Azure DevOps Pipelines and GitHub repositories while maintaining full governance and traceability across environments.

CHAPTER 7   REAL-WORLD CASE STUDIES AND FUTURE TRENDS

At its foundation, the pipeline starts from a **source control trigger**, typically a Git push or pull request to a shared **main** or **infra** branch. This event initiates the CI/CD pipeline, pulling from a version-controlled IaC repository structured around reusable Bicep or Terraform modules. The source repository is secured with **branch protection rules**, **signed commits**, and **CODEOWNERS** files to enforce collaborative discipline and reduce unauthorized changes.

The **Build Stage** includes pre-deployment actions like

- **Syntax validation** (e.g., bicep build or terraform validate)

- **Static analysis and linting** using tools like Checkov, TFLint, or PSRule

- **Security scanning** for secrets, misconfigurations, and CVEs using tools such as Trivy or Microsoft Defender for DevOps

- **What-If Analysis/Plan Phase**, generating a diff between the current state and desired configuration

This is followed by **automated unit tests** and **integration test hooks** (e.g., Terratest or Pester), which validate infrastructure logic without provisioning actual resources. The outcome of this stage is an **artifact** (e.g., compiled Bicep templates, plan files) that can be promoted through higher environments.

The **Release Stage** is broken into multiple environments: **Dev**, **QA**, **UAT**, and **Production,** each governed by environment-specific variables and secret scopes. Before promoting the artifact to each environment, the pipeline hits a **Policy Gate** integrated with

- **Azure Policy via REST API or extension tasks**, to evaluate compliance posture (e.g., disallowed SKUs, missing tags)

- **Open Policy Agent (OPA)/Gatekeeper or HashiCorp Sentinel**, when used with Terraform Enterprise or custom admission controllers

- **ServiceNow or Jira Approvals**, ensuring human intervention for sensitive changes

For critical environments like **Production**, **manual approval gates** are enforced. These approvals can be linked to designated approvers in Azure DevOps or GitHub environments, tied to business units, InfoSec, or compliance officers.

CHAPTER 7   REAL-WORLD CASE STUDIES AND FUTURE TRENDS

Each deployment step uses a **Service Connection** with least-privileged access (RBAC-scoped), authenticated via **Azure Workload Identity Federation (OIDC)** or **Managed Identity** to eliminate secret sprawl and credential management. All actions are logged in **Azure Monitor Logs**, GitHub Audit Logs, and optionally fed into **Microsoft Sentinel** for threat detection and change monitoring.

The pipeline is **declarative and reusable**, leveraging YAML-based templates, parameterized inputs, and conditional expressions to handle multi-region, multi-subscription, and tenant-specific deployments.

Additionally, **post-deployment gates** validate the resource state by

- Verifying drift detection using **what-if** or Terraform **plan** comparisons

- Checking runtime policies (e.g., NSG configurations, encryption status)

- Ensuring logs, metrics, and alerts are correctly wired into the observability stack

The final step includes **notifications and dashboard updates**, where success or failure messages are posted to Microsoft Teams, Slack, or email. Azure DevOps Pipelines can also emit telemetry to **Application Insights** or **Azure Dashboards** for centralized visibility.

In essence, **Figure 7-2** illustrates how secure CI/CD is not just about automation but about embedding policy-as-code, compliance controls, and operational awareness into every stage of the delivery pipeline. This model ensures that infrastructure changes are **repeatable**, **secure**, and **compliant by design**, aligning with both DevSecOps principles and enterprise regulatory frameworks.

Migration Workloads: From On-Premises to Azure

In large-scale enterprise transformations, the migration of critical workloads from on-premises infrastructure to Azure demands a structured, multiphase approach that balances modernization with operational continuity. This section outlines a real-world enterprise migration program involving heterogeneous systems, including databases, application servers, integration middleware, and legacy batch processes, successfully transitioned to a cloud-native architecture on Azure.

Workload Classification and Target State Mapping

The organization's on-premises landscape encompassed diverse platforms such as Oracle databases, .NET Core services hosted on IIS, COBOL batch jobs running on mainframes, and tightly coupled integration layers built on legacy enterprise service buses. To maximize Azure's platform-as-a-service (PaaS) and container-native capabilities, each workload type was mapped to its optimal Azure-native target state.

- **Relational Databases:** Oracle and SQL Server instances were assessed for compatibility and migrated to **Azure SQL Managed Instance**, which provided a managed environment with near 100% SQL Server compatibility, automatic patching, high availability, and native backup support.

- **Application Services:** Business-critical APIs and back-end services developed in **.NET Core** were containerized and deployed into **Azure Kubernetes Service (AKS)** for stateful workloads and into **Azure Container Apps (ACA)** for stateless microservices. This bifurcation allowed the team to optimize for scalability and fault isolation based on service behavior.

- **Integration Layers:** Monolithic integration components were re-architected into **event-driven architectures**, leveraging **Azure Event Grid** for reactive pub-sub messaging and **Azure Service Bus** for durable queue-based communication between services. This transition improved decoupling, observability, and integration latency.

- **Legacy Batch Systems:** COBOL-based batch processes were containerized using **App Service for Linux**, with execution support provided through **Windows Subsystem for Linux (WSL)**. This approach preserved core business logic while enabling DevOps automation and horizontal scalability.

Migration Strategy and Validation Practices

To minimize disruption and maintain production-grade confidence, each workload passed through a **dry-run execution** cycle in a dedicated **non-production subscription**. These dry runs included

- **Infrastructure Validation:** Deployment of IaC templates to mimic production topology.

- **Synthetic Monitoring Probes:** Health checks using **Application Insights** and **Azure Monitor** to assess runtime behavior.

- **Load Simulation:** Performance benchmarking using **JMeter** and custom traffic replays to measure latency, throughput, and scalability under stress conditions.

Post-migration, each workload underwent rigorous cutover rehearsals, rollback verification, and configuration hardening to align with the organization's security and compliance baselines.

Program Duration and Timeline

The complete migration program unfolded over a period of **14 months**, divided into four major phases:

1. **Discovery and Assessment (Months 1–3)**

   This phase involved comprehensive dependency mapping using Azure Migrate, business impact analysis, and cloud landing zone readiness. Workloads were categorized by criticality and technical complexity to prioritize sequencing.

2. **Core Workload Migration (Months 4–9)**

   High-impact databases and back-end services were transitioned first, enabling downstream systems to be reintegrated into the cloud environment. Azure DevOps pipelines were introduced to standardize deployment and post-deployment validations.

3. **Containerization and Modernization (Months 10–12)**

   Remaining legacy systems and internal tools were containerized, decoupled, and redeployed in AKS or ACA. CI/CD automation and policy-as-code frameworks were also integrated during this phase.

4. **Stabilization and Cutover (Months 13–14)**

    Final production cutovers were executed with rollback plans in place. Observability baselines were reviewed, and governance controls (e.g., tagging, cost alerts, Defender for Cloud policies) were applied uniformly across all migrated workloads.

The disciplined execution of this 14-month migration road map allowed the enterprise to transition mission-critical systems to Azure without business interruption while simultaneously modernizing its application stack, improving reliability, and reducing operational overhead.

Security, Risk, and Compliance Controls

The bank's internal audit team required audit logs for every cloud activity. Azure Policy was used to enforce diagnostic settings across all resources. Logs were centralized into a **Log Analytics Workspace per region** and retained for seven years in accordance with the data retention policy.

Azure Security Center (now Microsoft Defender for Cloud) was used for

- Threat detection across compute and network resources
- Continuous compliance scoring for CIS, NIST, and ISO 27001 benchmarks
- Alert forwarding to an integrated SIEM built on Azure Sentinel

Lessons Learned and Outcomes

Key Takeaways

- **Modular Landing Zones** enabled reuse across business units with minimal effort.
- **Centralized Policy Control** ensured governance without hampering developer agility.

- **High Observability** via Azure Monitor and Sentinel allowed proactive risk management.
- **Cost Optimization** was achieved through early commitment to Reserved Instances and shutting down dev environments nightly using Azure Automation Runbooks.

Business Impact

- 42% reduction in infrastructure TCO in the first 18 months.
- Developer environment provisioning dropped from ten days to under two hours.
- Improved audit scores in Singapore and Frankfurt, accelerating go-to-market for fintech APIs.

7.2 Case Study 2: Azure for National Health Systems—A Pandemic Response at Planetary Scale

When a national health authority was tasked with rolling out a centralized COVID-19 response platform during the early phases of the pandemic, the requirements were not just technically daunting; they were existential. Citizens needed timely test results, vaccination appointments, and accurate data on hospital capacity. Governments needed real-time analytics for policymaking. Researchers needed secure, anonymized datasets for epidemiological studies. Every stakeholder, from field nurse to federal administrator, depended on a digital backbone that didn't yet exist. This case study explores how Azure became the digital public health nervous system of an entire country, achieving hyper-scale agility, zero downtime, and trusted data flows, all while navigating one of the most turbulent socio-political environments in recent history.

The Urgency of Time and the Constraints of Policy

Unlike most digital transformation initiatives, the timelines for this project were set by biology and politics, not by procurement calendars or quarterly board meetings. The health system had less than 45 days to deliver:

- A **citizen-facing portal** for self-registration and vaccination scheduling

- A **clinical back end** integrated with 200+ hospitals, labs, and pharmacies

- A **real-time epidemiological dashboard** for command centers and policy units

- **Audit-grade traceability** and **data residency** for legal compliance

Legacy systems, many still hosted in on-prem data centers with nightly batch integrations, were fundamentally incompatible with such demands.

Azure was chosen not just for its cloud scale but for its **hybrid trust model**: integration with legacy EMR systems via on-prem ExpressRoute, layered with cloud-native apps that could scale elastically and push telemetry through APIs, Event Grid, and Azure Digital Twins.

Architecture Overview: A Nation-Sized Microservices Platform

The overall architecture reflected three operational planes: the citizen experience, the healthcare provider workflows, and the public health authority command tier. Azure services were orchestrated to reflect this separation of concerns while ensuring seamless data flow.

Citizen Portal Tier (Zone A)

- Deployed in Azure App Service across two paired regions (e.g., North Europe and West Europe)

- Fronted by Azure Front Door with WAF policies to handle DDoS and injection threats

- Utilized Azure Static Web Apps for public FAQs and content, with BFF (Backend for Frontend) APIs hosted in App Service (Linux)

CHAPTER 7 REAL-WORLD CASE STUDIES AND FUTURE TRENDS

Healthcare Provider Tier (Zone B)

- Hosted on AKS with .NET 6 microservices interfacing with HL7-based APIs for lab results, EMRs, and vaccine batch tracking.

- Azure API Management fronted these services to enforce throttling, transform headers, and validate schemas.

- Service Bus queues were used for order submissions, vaccine inventory updates, and asynchronous prescription workflows.

Public Health Command Tier (Zone C)

- Azure Synapse Analytics consumed telemetry from Event Hub.

- Data Lake Gen2 acted as a raw zone for storing longitudinal patient data.

- Power BI dashboards visualized regional test rates, ICU occupancy, and infection hotspots.

Each tier operated in separate subscriptions with different RBAC, policy, and logging scopes, enforcing *least privilege* while enabling federated operations.

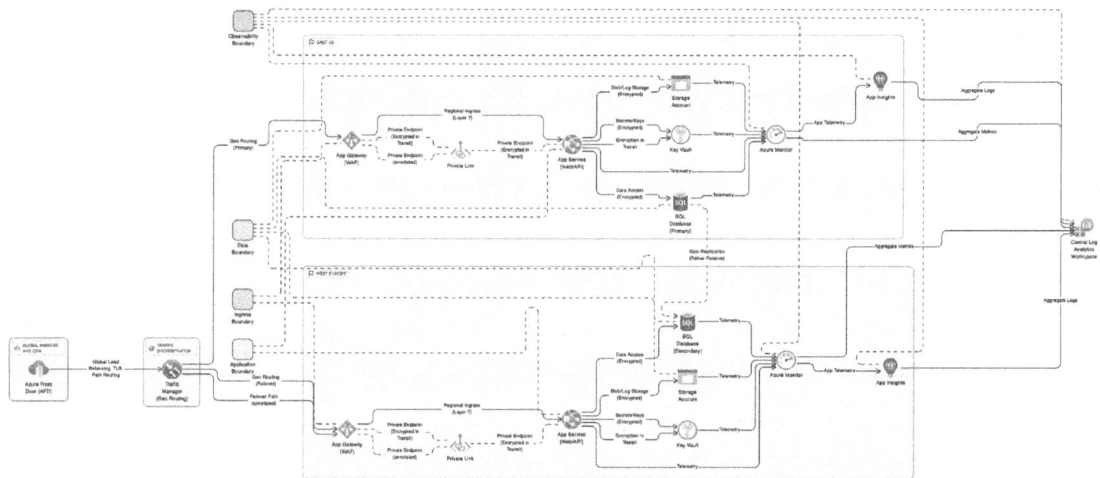

Figure 7-3. *Macro-architecture with Key Service Boundaries and Flows*

Figure 7-3 illustrates the macro-level architecture of an enterprise-grade Azure cloud environment, with clearly defined **service boundaries**, **data flow layers**, and **control plane integrations**. This architecture is designed to reflect a modular, scalable, and policy-compliant structure suitable for large organizations operating in regulated industries like finance, healthcare, or government.

At the core of this diagram lies the **Azure Landing Zone**, segmented into multiple **application zones**, **shared services**, and **platform components**. These logical boundaries map to separate **management groups**, **resource groups**, and **subscriptions**, providing a clear isolation of concerns and enabling RBAC (role-based access control) and policy scoping per service layer.

Starting from the **user access layer**, traffic originates from globally distributed clients, employees, APIs, and customers. These requests are initially processed through **Azure Front Door** or **Azure Application Gateway**, both of which offer SSL termination, Web Application Firewall (WAF) protection, and traffic routing based on geo-location, URL path, or latency.

The **API Management tier** (Azure API Management) acts as a secure façade and policy enforcement layer. This boundary mediates all north–south traffic, including identity federation via Azure AD, rate-limiting, and request inspection. API gateways route requests downstream to bounded application services.

Each **Application Service Boundary** for domains like Billing, Customer Management, or Analytics is encapsulated into independent Kubernetes namespaces (AKS), Azure App Services, or Function Apps. These are versioned, CI/CD deployed, and integrated with **Azure Monitor**, **Application Insights**, and **Log Analytics** workspaces.

For **stateful data processing**, services interact with dedicated data stores located in the **Data Boundary**. This tier includes Azure SQL Database for structured data, Azure Cosmos DB for globally distributed NoSQL workloads, and Azure Data Lake for raw ingestion and archive storage. Data movement across layers is governed via **Event Grid**, **Service Bus**, and **Azure Data Factory**, ensuring asynchronous and decoupled processing patterns.

In parallel, a **Control Plane Layer** ensures compliance and governance. Azure Policy is scoped at the management group level, enforcing guardrails such as encryption-at-rest, tagging standards, and network boundaries. Microsoft Defender for Cloud provides posture management and threat detection across compute, data, and identity planes.

Supporting these functional services is a **Shared Services Layer** hosting core utilities like

- Centralized Azure Key Vaults
- DNS zones
- Azure Bastion and Jumpboxes
- Update management and patch automation
- Container Registry (ACR) and GitHub Enterprise

Integration across these boundaries is conducted via **managed identities**, **private endpoints**, and **Azure Private Link**, ensuring secure traffic flow with minimal exposure to the public internet.

Operational flows shown as directional arrows demonstrate how requests travel from end users to compute resources, through API gateways and firewalls, down to back-end services and data platforms. Outbound telemetry and logs flow toward a **central observability stack**, while audit trails feed into compliance dashboards powered by **Azure Workbooks**, **Sentinel**, or **third-party SIEMs**.

In summary, Figure 7-3 presents a layered enterprise macro-architecture that cleanly separates **business domains**, **operational services**, **security boundaries**, and **governance mechanisms**. This design allows for parallel team development, simplified change control, Zero Trust enforcement, and accelerated deployment cycles without compromising on compliance or performance.

Secure Identity Federation and Pseudonymized Data Design

In a health context, identity is more than a login credential; it's a chain of custody over patient data. Azure AD B2C was configured for citizens with multi-factor support (via SMS, authenticator app, and device trust). Health professionals used Azure AD (federated with HIE's ADFS) with strict Conditional Access policies:

- Device compliance
- Geo-fencing (e.g., no logins from outside national boundaries)
- Role elevation using Privileged Identity Management (PIM)

To ensure research usability while preserving privacy, **patient records were pseudonymized** using Data Factory transformations with encryption at rest and in transit. Data masking functions ensured identifiers were not exposed in analytics views. Azure Confidential Ledger was used to maintain tamper-proof audit trails of data access for legal review and breach response compliance.

Integration with Legacy Systems via Hybrid Models

Many public hospitals and laboratories still operate legacy systems, some over 15 years old, with only CSV or SOAP interfaces. Azure Integration Services (Logic Apps, API Management, Hybrid Connections) were pivotal.

- **Logic Apps** polled and transformed FTP-based lab reports every 15 minutes.
- **Hybrid Connections** tunneled securely into on-prem data centers without requiring full VPN exposure.
- **Azure Arc** was used to register on-prem SQL servers and Kubernetes clusters, enabling centralized policy enforcement and monitoring through Azure Security Center.

This hybrid model enabled cloud-speed innovation without waiting for on-prem modernization, a classic *strangler fig* pattern in action.

Observability and Resilience Engineering

Given the project's political visibility, **downtime was not an option**. Azure Monitor, Application Insights, and Log Analytics were integrated across all tiers. Custom dashboards tracked

- API latency trends
- Geo-spread of traffic
- Exception counts per microservice
- Retry rates in queues and Event Hubs

Resilience engineering principles were embedded at design time:

- AKS horizontal pod autoscalers and disruption budgets
- Geo-redundant storage with RA-GRS for critical patient records
- Global distribution of App Configuration and Key Vault using Azure's multi-region features

Synthetic transaction scripts are executed every five minutes to simulate real-world user behavior and trigger alerts on anomalies.

CHAPTER 7   REAL-WORLD CASE STUDIES AND FUTURE TRENDS

DevOps at Pandemic Speed

The urgency of the initiative required **daily production pushes,** a scenario unimaginable in most healthcare IT environments. GitHub Actions pipelines were used with environment promotion gates. Key practices included

- Infrastructure defined in Bicep, versioned and validated through pre-merge checks

- App binaries scanned via Microsoft Defender for DevOps before image build

- Canary deployments using Azure App Service deployment slots

- Chaos testing conducted using Azure Chaos Studio to simulate latency, memory leaks, and failover

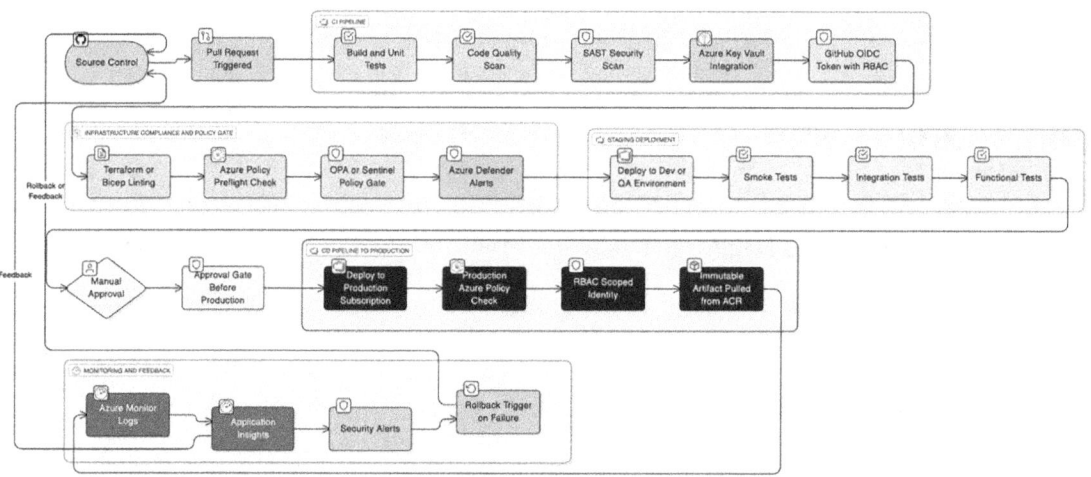

Figure 7-4. *CI/CD Pipeline with All Compliance and Approval Checkpoints*

Figure 7-4 presents a modern, enterprise-grade Continuous Integration and Continuous Deployment (CI/CD) pipeline designed for Infrastructure as Code (IaC) and application workloads on Microsoft Azure. This architecture integrates not only build and release stages but also embeds **policy-as-code**, **compliance enforcement**, and **manual approval gates** as first-class citizens throughout the software delivery life cycle.

At the leftmost stage of the pipeline lies the **Source Code Repository**, typically hosted in **Azure Repos** or **GitHub Enterprise**, where IaC modules (Terraform, Bicep), application code (e.g., .NET, Python, Java), and reusable pipeline templates are maintained. Branch policies enforce peer-reviewed pull requests (PRs), code signing, and commit hygiene checks before merging into the mainline (e.g., main, sharerg, or release branches).

Once a merge occurs, the pipeline is automatically triggered, moving to the **Build Stage**. Here, **static code analysis** tools such as **tflint**, **checkov**, or **bicep linter** are executed to catch early misconfigurations and security issues. Simultaneously, **unit tests**, **format validation**, and **schema enforcement** validate both the application logic and IaC templates.

The next phase, the **Policy Validation Stage**, is critical in regulated environments. In regulated environments, enforcing compliance *before* deployment is critical. To achieve this, infrastructure pipelines are designed to invoke **policy-as-code frameworks** such as **Azure Policy**, **Terraform Sentinel**, or **Open Policy Agent (OPA)** during the validation phase. While OPA is commonly associated with Kubernetes admission control via Gatekeeper, it is not limited to that use case. In this context, OPA is integrated directly into the CI/CD pipeline (e.g., GitHub Actions or Azure DevOps) as a **pre-deployment check**. Resource definitions such as Terraform plans or ARM/Bicep templates are evaluated against custom OPA policies to ensure they meet preapproved security, compliance, and tagging standards **before** they are applied to any Azure environment.

Examples include

- Ensuring diagnostic logs are enabled for all resources
- Blocking public IP exposure for PaaS services
- Verifying tags like **costcenter**, **env**, and **owner** are present

If policy validation fails, the pipeline is halted with a detailed compliance report, ensuring that only compliant infrastructure moves forward.

Following this, the **Plan and Review Stage** leverages tools such as terraform plan or az deployment what-if to produce a **change set**, which is then reviewed by senior architects or change approvers. This output is either posted as PR comments (in GitHub) or pushed to **Azure Boards or ServiceNow** for traceability. This stage is enhanced by **manual approval gates** that prevent automatic progression to deployment, especially critical for production workloads.

Once approved, the pipeline proceeds to the **Deploy Stage**, which often uses a **matrix strategy** or environment-based scoping. Deployments are gated per environment (e.g., dev, qa, prod) and use parameterized templates. Each environment may include additional post-deployment checks:

- **Smoke Tests** using curl, Postman, or Playwright
- **Post-deploy compliance scans** with Azure Defender and Microsoft Purview
- **Backout validation** via health probes and auto-remediation scripts

The final phase is **Monitoring and Feedback**, where all telemetry from the deployment (success/failure, duration, compliance drift) is collected using

- **Azure Monitor and Log Analytics**.
- **Custom dashboards and KPIs** in Azure Workbooks
- **Notifications** through Teams, Slack, or ITSM systems

In parallel, **pipeline artifacts**, audit logs, and plan results are versioned and stored in **Azure Artifacts**, **GitHub Releases**, or **Blob Storage** for traceability and rollback.

This pipeline model not only ensures **secure and repeatable deployments** but also aligns with the principles of **DevSecOps** and **Compliance-as-Code**. By tightly coupling automation with policy enforcement, human approvals, and post-deployment verification, the organization can meet **SOX**, **HIPAA**, **PCI-DSS**, or **ISO 27001** requirements without slowing down innovation.

In essence, Figure 7-4 visualizes the convergence of **engineering velocity** and **compliance rigor**, forming the backbone of a production-ready Azure CI/CD strategy in complex enterprise environments.

Governance, Audit, and Compliance Assurance

The national data protection authority required daily exports of system logs and user access events. Azure Policy was used to enforce

- Encryption at rest and in transit
- Diagnostic settings on every resource (e.g., Cosmos DB, Key Vault, App Services)
- Deny policies for public IP exposure

Azure Purview (now Microsoft Purview) cataloged sensitive data and tracked lineage from ingestion to report. Every user action API call, data export, and login was logged into an immutable audit store with seven-year retention.

Outcomes and Transformational Impact

- Over **36 million citizens onboarded** within the first 90 days
- **< 1.2s average API response time**, even during traffic surges of over five million hits per hour
- **Zero production downtime** across two years, validated by public SLA reports
- Enabled real-time, regional policy decisions with dashboard latency under 60 seconds

But beyond the numbers, the true success lay in the **digital trust** built with citizens and the **paradigm shift** in how governments could think about agility, privacy, and resilience powered by the cloud.

7.3 Case Study 3: Azure for Ecommerce Scale—A Global Retail Platform's Peak Readiness Strategy

As ecommerce matured from a convenience to a necessity, consumer expectations evolved into unforgiving benchmarks: 24/7 availability, sub-second page loads, and personalized shopping experiences across every device and geography. Nowhere is this demand more intense than during peak seasons, Black Friday, Cyber Monday, Diwali, and Singles' Day, when traffic surges 10× to 20× normal levels within minutes. This case study explores how a global retail conglomerate with multiple brand storefronts built a modern, microservices-based commerce platform on Azure, capable of scaling predictably, delivering with speed, and maintaining ironclad reliability under extreme pressure.

From Monoliths to Microservices: A Mandate from the CTO

The previous architecture was a traditional ecommerce monolith: a Java-based application running on a commercial app server, backed by a single RDBMS cluster in a co-located data center. The system was rigid, costly to scale, and riddled with single points of failure. The CTO's directive was unambiguous:

> "We must decouple to survive the next five years. Each domain must be able to evolve, scale, and recover independently."

The architecture transformation adopted a **Domain-Driven Design (DDD)** approach, breaking down the platform into 14 core domains: Catalog, Inventory, Pricing, Promotions, Cart, Checkout, Payments, Order Management, Shipping, Customer Identity, Loyalty, Reviews, Notifications, and Content.

Each domain became a microservice, implemented in .NET 7 or Node.js, depending on the team's preference. Every microservice owned its data, exposed APIs via Azure API Management, and communicated via asynchronous messages using Azure Event Grid and Service Bus.

Azure Architecture for Scale, Speed, and Safety

The entire solution was deployed across **two Azure regions** (East US and West Europe), configured in **active-active mode** using Azure Front Door for global load balancing and traffic shaping.

Core Infrastructure Design

- **Microservices:** Deployed as Docker containers in Azure Kubernetes Service (AKS) clusters across both regions, scaled using Horizontal Pod Autoscalers (HPA) based on CPU and queue length
- **Data Services**
  - Azure Cosmos DB (with multi-region writes) for Cart, Orders, and Inventory
  - Azure SQL Database Hyperscale for transactional services like Payments and Loyalty

- Redis Enterprise on Azure for caching product details, session data, and promo rules

- **Content Delivery**: Static assets and images served via Azure CDN with regional Points of Presence (PoPs), integrated with Azure Storage for content origin

- **Telemetry and Logs**: Centralized in Azure Monitor and Application Insights with custom KQL dashboards tracking shopping funnel metrics (home ➤ product ➤ cart ➤ checkout)

CHAPTER 7 REAL-WORLD CASE STUDIES AND FUTURE TRENDS

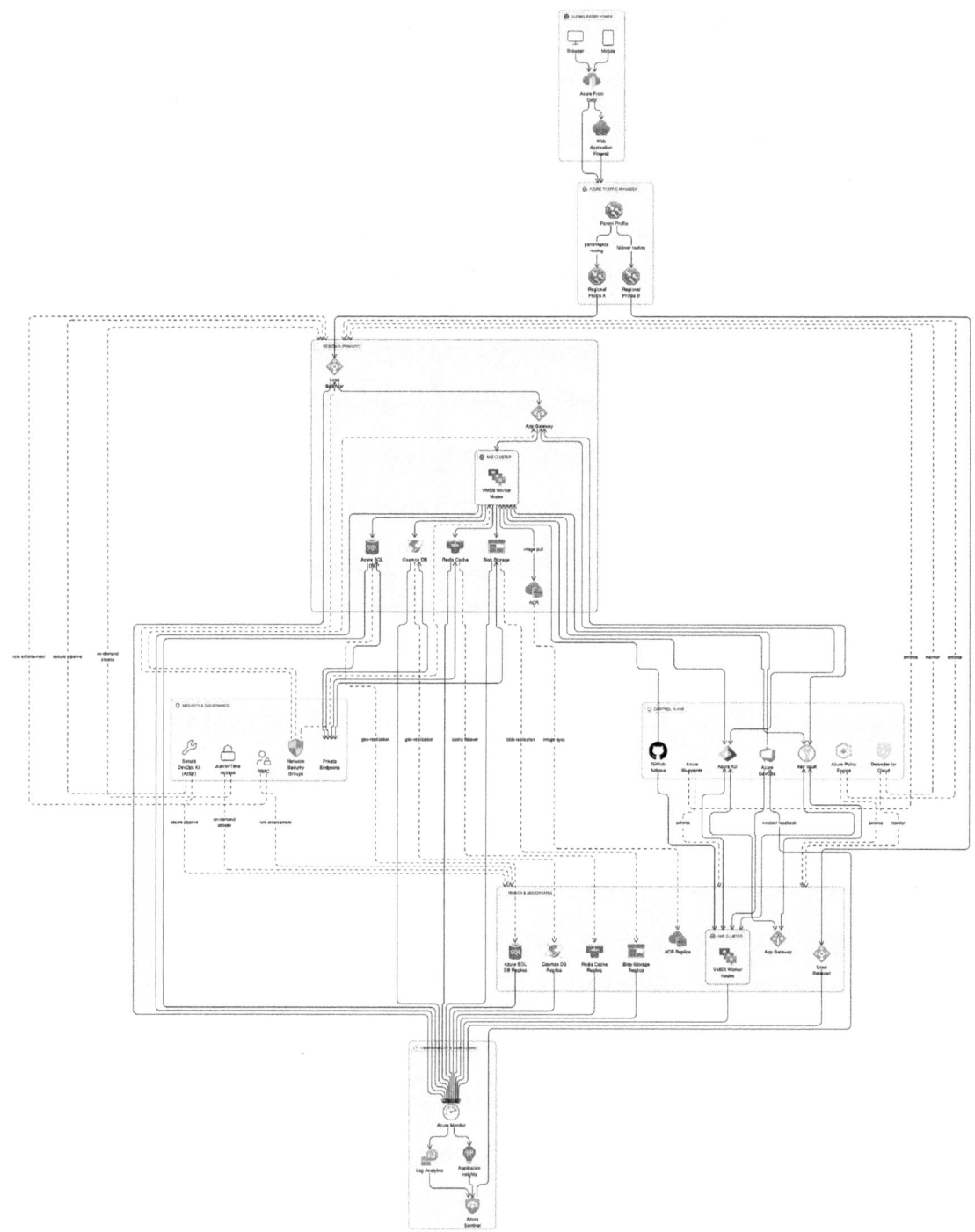

Figure 7-5. *Complete Topology, Including Data Paths, Failover Zones, and Control Surfaces*

CHAPTER 7   REAL-WORLD CASE STUDIES AND FUTURE TRENDS

Figure 7-5 presents a comprehensive architectural blueprint of an enterprise-grade Azure deployment engineered for resiliency, performance, and governance. This topology integrates data paths, control planes, and failover zones to ensure uninterrupted operations, secure access, and real-time observability across global workloads.

At the heart of the topology lies the **Hub-and-Spoke network architecture**, deployed across **two Azure regions, primary** and **secondary (failover),** with seamless integration via **Azure Global VNet Peering** and **Private Link**. The **hub network** in each region hosts shared resources such as **Azure Firewall, DNS forwarders, Azure Bastion,** and **Network Virtual Appliances (NVAs)**, acting as a centralized control zone. These hubs are connected to on-premises networks through **ExpressRoute circuits**, with **dual connectivity** for carrier redundancy.

Each **spoke network** is dedicated to an application domain or workload tier: microservices, data services, user-facing applications, and background jobs. These are deployed using **Azure Kubernetes Service (AKS), App Services,** or **VM Scale Sets**, depending on SLA and compute requirements. Traffic between application tiers is regulated using **NSGs, UDRs,** and **Azure Firewall policies**, with identity-aware microsegmentation enforced via **Azure AD Conditional Access** and **Azure Policy**.

The **data path layer** is engineered for both real-time and batch flows. Event-based workloads leverage **Azure Event Hubs, Azure Service Bus,** or **Kafka on HDInsight**, while batch data pipelines utilize **Azure Data Factory** and **Synapse Pipelines**. All storage, whether structured, unstructured, or analytical, is protected via **Customer-Managed Keys (CMK), Private Endpoints,** and **immutable blob policies**. **Data encryption in transit and at rest** is strictly enforced using **TLS 1.2+** and **Azure Defender for Storage**.

To ensure fault tolerance, the topology integrates **Traffic Manager with Nested Profiles**, directing user requests to the most performant regional endpoint or automatically failing over during regional outages. In parallel, **Azure Front Door** handles global HTTP/HTTPS routing with Web Application Firewall (WAF) capabilities and dynamic site acceleration (DSA). Back-end application availability is reinforced by deploying across **availability zones** and **availability sets**, ensuring resiliency at both the data center and regional levels.

The **control surfaces** are meticulously isolated. Infrastructure provisioning is driven via **Azure DevOps Pipelines** and **Terraform Enterprise**, using service principals and managed identities scoped with least-privilege roles. **Azure Blueprints** and **custom policy initiatives** ensure compliance with organizational and regulatory baselines such as ISO 27001, HIPAA, and CIS benchmarks.

Operational visibility is embedded via the **Observability Stack** comprising:

- **Azure Monitor** for platform metrics
- **Log Analytics** for unified telemetry and queryable logs
- **Application Insights** for end-to-end distributed tracing
- **Azure Workbooks and Dashboards** for real-time KPIs and compliance scorecards

Each component in this topology contributes to a cohesive system designed for **operational excellence**. From CI/CD-integrated policy enforcement and fine-grained network segmentation to encrypted telemetry and scalable failover, this architecture enables enterprises to meet stringent SLAs, security mandates, and operational objectives across regions and business units.

This figure serves as the reference implementation for readers seeking to design or benchmark high-assurance, mission-critical Azure deployments in regulated or performance-sensitive environments.

Real-Time Personalization with Azure AI

Personalization was central to the experience recommendation engines, dynamic pricing, and behavioral targeting were all driven by customer context. Azure Machine Learning services were used to train models on user data piped from

- Azure Synapse (clickstream analytics)
- Azure Data Lake (purchase history and inventory)
- Customer segments curated in Microsoft Customer Insights

Personalized product recommendations were served via REST APIs that ran on Azure Kubernetes Services and were fronted by Azure Front Door Rules Engine to conditionally cache based on user cohort. Training jobs were orchestrated via Azure Machine Learning pipelines with auto-scaling compute clusters for efficient batch learning.

Observability for Peak Readiness

The company engineered **Peak Readiness Playbooks** that relied heavily on proactive observability. These included

- **Load profiles simulated with Azure Load Testing** based on previous years' traffic patterns.

- **Synthetic users emulated via Azure Chaos Studio** to validate regional failovers and ensure DNS, front-end services, and data pipelines were resilient to node failures or regional outages.

- **Red-Green dashboards in Azure Workbook**, where each microservice displayed real-time SLA compliance, error budgets, and user-perceived latency broken down by geo.

An **SRE (Site Reliability Engineering)** team ran daily "health checks" and monthly game days scenarios where entire regions were failed over or specific services were artificially degraded to test alert fidelity and recovery paths.

Infrastructure as Code and Deployment Strategy

Each service's infrastructure was codified in **Terraform**, stored in GitHub Enterprise with GitOps pipelines. A central team maintained "Golden Modules" for

- AKS clusters with guardrails (network policies, admission controllers, Prometheus sidecars)

- Azure Front Door rulesets and WAF policies

- Azure SQL with geo-failover groups and managed identity wiring

- Logging policies and Diagnostic Settings enforcement via Azure Policy

Deployment Pipelines used GitHub Actions with the following principles:

- **Preview environments** spun up automatically via pull requests

- **Canary rollouts** using AKS blue-green deployments

- **Time-windowed deployments** to limit risk exposure during peak hours (e.g., 1:00 AM–4:00 AM UTC)

All images were stored in Azure Container Registry with per-microservice base image hardening (e.g., Alpine or Distroless), vulnerability scans, and version pinning.

Business Impact

The new Azure-powered platform launched ahead of the holiday season. The results were dramatic:

- **99.995% uptime** across both regions, despite traffic doubling YoY
- Average **checkout latency of <800ms**, even during peak sale events
- Zero service disruptions during Black Friday, supported by dynamic autoscaling
- **300+ deployments to production per week** with no regression-induced outages

The platform also gave the company the agility to launch new brands and market segments, each as isolated microservices without impacting the core system.

7.4 Case Study 4: Financial Services on Azure—Building a Real-Time Risk Engine for Global Trading

The Mission-Critical Imperative of Risk at Millisecond Speed

For capital markets institutions, risk isn't reviewed quarterly; it's calculated every microsecond. As market conditions shift and trades are executed in real time, the value-at-risk (VaR), exposure, and margin obligations of a firm evolve dynamically. In the wake of new regulations under Basel III, MiFID II, and Dodd-Frank, global banks have been mandated to implement near-real-time risk visibility across asset classes and jurisdictions.

This case study explores how one of the world's top investment banks re-architected its risk analytics platform from a batch-driven, overnight batch system into a **low-latency, real-time engine** deployed across multiple Azure regions. Their primary goal:

to achieve **intraday risk aggregation** across 120+ trading desks while maintaining regulatory compliance, system resilience, and performance guarantees under 30 milliseconds.

Legacy Bottlenecks and Azure Motivation

The bank's prior architecture ran on-premises in regional data centers. End-of-day trades were batch processed using a monolithic Java application and loaded into a massive Oracle database. Latency was measured in hours, not milliseconds. The critical issues included

- **No Real-Time Capability**: Risk teams lacked up-to-date exposure calculations during trading hours.

- **Hardware Constraints**: On-prem clusters couldn't scale fast enough during volatility spikes.

- **Compliance Challenges**: Regulatory requirements demanded audit trails, access logging, and disaster recovery, not easily enforced in the legacy stack.

The technology transformation was driven by the CTO's strategic alignment to Azure due to its **regulated cloud certifications (SOC, ISO, PCI-DSS, FedRAMP, etc.)**, **financial services compliance blueprints**, and the ability to build an **event-driven, streaming analytics architecture** that scales horizontally.

Architecture Overview: A Streaming-Based Risk Engine on Azure

The new system used **Apache Kafka for ingestion**, **Azure Event Hubs for streaming**, and **Azure Databricks for real-time analytics**. Each trade generated by upstream trading platforms (Equities, FX, Derivatives, Fixed Income) was immediately pushed into Azure via Kafka MirrorMaker 2, then funneled into Event Hubs.

Risk Calculation Pipelines were implemented as follows:

- **Ingestion Tier**: Kafka topics mirrored into Azure Event Hubs

- **Streaming Aggregators:** Azure Databricks notebooks using Spark Structured Streaming, performing joins across reference data (from Azure SQL) and market data feeds (via Azure Data Explorer)

CHAPTER 7 REAL-WORLD CASE STUDIES AND FUTURE TRENDS

- **Model Execution**: VaR, Greeks, and Exposure models executed as ML pipelines deployed in Azure Kubernetes Service

- **Intermediate Caching**: Azure Cache for Redis used for pre-aggregated instrument-level metrics

- **Final Aggregation and Storage**: Daily and intraday summaries written to Azure Synapse Analytics and archived in Azure Data Lake Gen2 for compliance retention

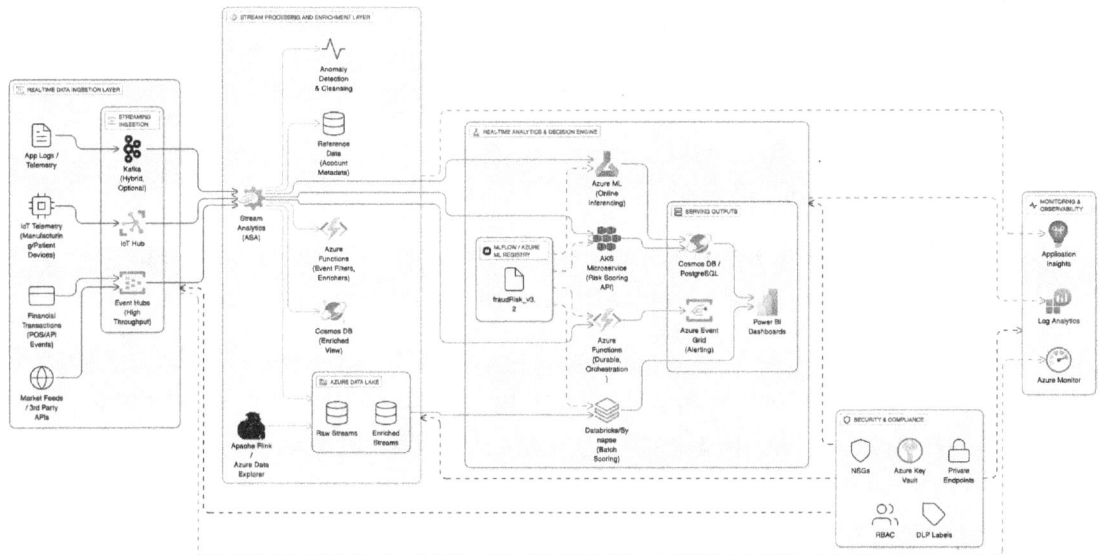

Figure 7-6. *Complete Streaming Topology and Risk Analytics Engine Across Ingestion, Transformation, and Model Execution Tiers*

Figure 7-6 depicts the complete streaming architecture for a real-time fraud detection and risk analytics solution built on Azure. This topology spans multiple architectural layers, orchestrating ingestion, enrichment, inference, and visualization using a robust set of Azure-native services.

On the left, the **Real-Time Data Ingestion Layer** ingests diverse data streams from operational systems. This includes logs and telemetry from applications, IoT data from manufacturing and patient monitoring devices, market feeds and third-party APIs, and high-throughput financial transactions (such as POS and API-based payment events). These are funneled through Azure Event Hubs and IoT Hub, with optional support for Kafka as a hybrid ingress mechanism.

Moving to the **Stream Processing and Enrichment Layer**, Azure Stream Analytics (ASA) serves as the central engine. It applies filters, joins, and transformations by referencing contextual metadata (like account profiles) and detecting anomalies in-flight. Output streams are bifurcated: one path flows to **Azure Data Lake** (for raw and enriched stream persistence) and another feeds downstream analytics.

The **Real-Time Analytics and Decision Engine** leverages multiple compute modalities. For low-latency inference, Azure Machine Learning (Azure ML) serves models trained and versioned via MLFlow or the Azure ML registry. A containerized microservice hosted in AKS (Azure Kubernetes Service) acts as the online scoring API, orchestrated by Azure Functions when durability or stateful orchestration is required. For deeper, non-real-time insights, batch scoring is handled by Azure Databricks or Synapse Analytics.

In the **Serving Outputs Layer**, risk scores and alerts are persisted in Cosmos DB or PostgreSQL and published to Azure Event Grid for downstream consumers. Business stakeholders and fraud analysts consume insights via Power BI dashboards updated in near real time.

The **Monitoring and Observability Layer** integrates Application Insights, Log Analytics, and Azure Monitor to provide full-stack visibility across microservices, data pipelines, and analytics components. This observability fabric ensures health, performance, and traceability across the entire life cycle.

Finally, **Security and Compliance Controls** are embedded at every layer. These include Azure-native NSGs (Network Security Groups), RBAC (role-based access control), private endpoints, DLP labels (Data Loss Prevention), and secrets managed via Azure Key Vault. Together, these ensure that data remains protected, compliant, and accessible only through governed interfaces.

In summary, Figure 7-6 illustrates a highly cohesive, production-grade architecture for streaming analytics, where ingestion pipelines, machine learning, and secure, observable infrastructure converge to deliver high-confidence, low-latency decisions across industries like finance, healthcare, and manufacturing.

Zero Trust, High Compliance Architecture

Given the financial and reputational risks, the architecture enforced **Zero Trust principles** and followed Microsoft's **Cloud Adoption Framework for Financial Services**:

- **Network Isolation**: AKS nodes were deployed in private subnets with NSGs locking access to specific service endpoints. No public ingress.

- **Data Sovereignty**: Regional data residency enforced via paired Azure regions (e.g., UK South ↔ UK West; East US ↔ Central US).

- **Key Vault + HSM**: All secrets managed via Azure Key Vault backed by hardware security modules (HSMs) meeting FIPS 140-2 Level 3 compliance.

- **Immutable Audit Trails**: Event Hubs Capture was used to archive all trade events in Azure Data Lake for non-repudiation.

- **Policy Enforcement**: Azure Policy ensured encryption-in-transit/at-rest, diagnostic settings, and storage replication policies.

Identity was federated using Azure AD B2B with Conditional Access and Privileged Identity Management (PIM) for all admin access.

Performance Engineering for Millisecond Analytics

To meet the sub-30ms SLA, engineering teams employed a combination of

- **Azure Proximity Placement Groups**: Ensured compute and storage nodes were deployed in low-latency network planes.

- **.NET 8 + Native AOT**: Performance-critical services (model runners) were implemented using Native AOT compilation for faster cold starts.

- **High-Throughput SKUs**: Azure Event Hubs Premium Tier with 100+ throughput units to avoid throttling during market volatility.

- **AKS Node Pools**: Tiers of node pools (GPUs for Monte Carlo simulations, CPU-optimized pools for streaming) to match workload profiles.

Monitoring was performed using Azure Monitor with real-time latency histograms, correlated with model output times and Kafka lag metrics.

Continuous Deployment with Guardrails

Infrastructure was managed via **Bicep** templates, versioned in Azure DevOps Repos, and deployed using **multistage YAML pipelines** with built-in approval gates. Every microservice followed

- **Blue-Green deployment strategy** with readiness and liveness probes
- **Release Rings**: Canary environments before production rollout
- **Chaos Engineering** using Azure Chaos Studio: Simulated latency, node failures, and pod evictions during trading hours under controlled conditions

Business Outcome and Regulatory Confidence

Since go-live:

- The system processes **over 100 million trade events per day**, with 99.98% of model outputs delivered within 30ms.
- Compliance teams gained **real-time dashboards** showing exposure by desk, asset class, and region.
- Regulatory audits passed without major findings due to automated evidence generation from Azure Policy, Monitor Logs, and Compliance Manager.
- Data scientists could **train and retrain risk models** daily, integrating new variables (ESG scores, credit risk flags) without production downtime.

7.5 Case Study 5: Healthcare on Azure—Modernizing Electronic Health Records (EHR) with FHIR APIs and Data Interoperability

The Healthcare Mandate: From Siloed Systems to Interoperable Care

Healthcare is a domain riddled with complexity, not only because of its regulatory sensitivities but also due to the fragmented nature of legacy systems. In most countries, Electronic Health Record (EHR) systems evolved in isolation across providers, hospitals, and insurers, leading to a landscape where

- Patient records were siloed across geographies and organizations.
- Interoperability between systems was minimal or nonexistent.
- Data exchange was often manual, non-standardized, and prone to delay or error.
- Providers lacked a unified view of a patient's history, undermining care quality.

This case study centers on a national health service provider that embarked on a multiyear digital transformation to implement **FHIR (Fast Healthcare Interoperability Resources)** APIs and achieve **seamless data exchange across hundreds of hospitals**, using Azure as the secure cloud foundation.

Strategic Vision: FHIR-First, API-Enabled, Cloud-Native

Guided by mandates such as the U.S. ONC 21st Century Cures Act and Europe's EHDS initiative, the organization chose Azure for three reasons:

1. **Native support for HL7 FHIR via Azure API for FHIR** offering out-of-the-box compliance with data formats, security, and SMART on FHIR authorization
2. **Global scale with sovereign cloud options**, supporting regional residency and GDPR compliance

3. **Deep integration with analytics, machine learning, and AI services**, to enable future use cases such as clinical decision support and population health analytics

The architecture was centered around a **FHIR Data Hub** deployed on Azure, enabling ingestion, validation, storage, and secure API exposure of patient data from disparate hospital systems.

Reference Architecture: Azure FHIR Hub with Smart App Integration

The architecture featured the following components:

- **FHIR Server**: Azure API for FHIR hosted in a private subnet using Azure Private Link and integrated with Azure AD B2C for patient and practitioner access control

- **Ingestion Layer**: Logic Apps and Azure Functions to parse, transform, and validate incoming HL7 v2 messages into FHIR format

- **EHR Integration Agents**: Deployed as containerized apps in on-premises hospitals, publishing data securely to Azure Event Grid and API endpoints

- **Data Lake Gen2 Archive**: Raw and normalized data persisted for auditing and reprocessing needs

- **Power BI Dashboards**: Used by healthcare administrators to view operational metrics such as admissions, discharges, and emergency alerts.

SMART on FHIR apps accessed the FHIR APIs via OAuth 2.0 flows with fine-grained scopes, enforcing data minimization per clinical role.

CHAPTER 7   REAL-WORLD CASE STUDIES AND FUTURE TRENDS

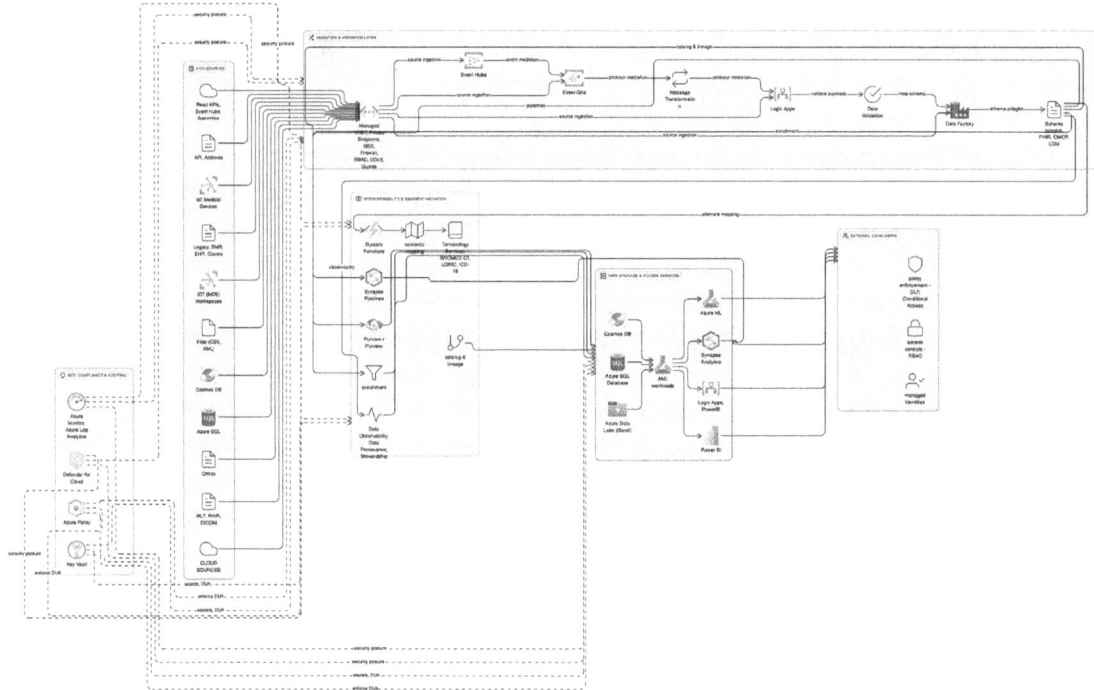

Figure 7-7. *End-to-End Data Interoperability Architecture Powered by Azure Services*

Figure 7-7 illustrates a fully integrated Azure data interoperability architecture that enables seamless data flow across ingestion, processing, storage, analytics, and consumption layers within a highly governed enterprise data ecosystem. This architecture is designed to support large-scale, multisource data ingestion, cross-domain harmonization, and standardized access patterns for both operational and analytical workloads.

At the **ingestion layer**, diverse data sources, including enterprise SaaS platforms (like Salesforce, Dynamics 365), on-premises systems (via Azure Data Gateway or ExpressRoute), and IoT edge devices, are integrated using **Azure Data Factory**, **Event Hubs**, and **IoT Hub**, depending on the nature and velocity of the data. Batch and real-time streams are simultaneously supported, with built-in schema registration and versioning via **Azure Schema Registry**.

The **transformation and processing tier** leverages **Azure Synapse Pipelines**, **Data Flows**, and **Azure Databricks** notebooks for scalable ETL/ELT operations. Standardized transformation logic ensures semantic consistency across business domains through the use of canonical models and mapping templates. Workflows are orchestrated using **Azure Logic Apps** and monitored centrally via **Azure Monitor** and **Log Analytics**.

All transformed datasets are stored in **Azure Data Lake Storage Gen2** using a medallion architecture (Bronze, Silver, Gold zones) that facilitates auditability, traceability, and multi-tenant data separation. This serves as the **interoperable data layer**, governed by **Azure Purview** for cataloging, classification, and lineage tracking. Data contracts and access policies are enforced through **Azure RBAC**, **Microsoft Entra ID**, and **Managed Identities**.

For analytical workloads, **Power BI**, **Azure Synapse Analytics**, and **Azure Machine Learning** integrate directly with curated datasets, offering users governed self-service analytics and ML model access. Simultaneously, operational APIs and downstream systems consume data via **Azure API Management**, **Azure Data Share**, or **OData endpoints**, ensuring consistency across business functions.

Cross-cutting concerns such as data quality, schema evolution, SLA monitoring, and privacy compliance (GDPR, HIPAA) are addressed via embedded controls throughout the pipeline, making the architecture robust, enterprise-ready, and audit-friendly.

Ultimately, this architecture empowers organizations to achieve true data interoperability across hybrid estates, enabling real-time insights, scalable machine learning, and a data-driven culture while remaining compliant with internal and external governance mandates.

Ensuring Privacy and Compliance: HIPAA, GDPR, and Beyond

The solution was architected under a **compliance-by-design** model. Key controls included

- **Encryption**: All data in transit and at rest used AES-256. Azure Disk Encryption with customer-managed keys (CMKs) enforced regulatory compliance.

- **Role-Based Access Control (RBAC)**: Practitioners, data analysts, and researchers were assigned RBAC roles at the resource group level and scoped to patient cohorts via Azure Purview.

- **Consent Management**: Patient consents were recorded via a dedicated module and enforced programmatically before granting data access.

- **Immutable Logging**: Azure Monitor, Log Analytics, and Diagnostic Settings captured every API call, including payload hash, timestamps, and client IPs.

Quarterly audits validated that **PHI (Protected Health Information)** flows were isolated from analytics workloads and that access patterns conformed to defined policies.

DevOps for Digital Health: From Sandbox to National Rollout

Infrastructure as Code (IaC) and DevOps played a pivotal role:

- **Bicep Modules** were authored for FHIR API deployment, firewall rules, identity configuration, and diagnostic settings.

- **CI/CD Pipelines** in GitHub Actions validated API schema compliance, ran synthetic tests, and deployed into regional environments.

- **Release Stages**: Sandbox ➤ Pre-Prod ➤ Prod, with policy-based approvals and anomaly detection gates.

A secure build pipeline also used **Azure Container Registry (ACR)** with immutable image tags for containerized EHR adapters. These adapters implemented transformation logic using .NET and were rolled out to each hospital's on-prem Kubernetes cluster.

Outcomes: Healthier Ecosystems, Healthier Patients

The modernization delivered tangible clinical and operational impact:

- **Reduction in Duplicate Tests and Medication Errors**: Shared allergy, medication, and diagnostic data across regions.

- **Faster Admissions and Discharges**: Automating transfer of patient history across hospital systems.

- **Enhanced Research Insights**: De-identified datasets were made available for epidemiology studies using Azure Synapse and ML models in Azure Machine Learning.

- **Higher Patient Engagement**: Patients could access their records and share them with specialists using FHIR-compatible mobile apps.

Regulators hailed the solution as a blueprint for **cloud-native interoperability in public healthcare systems**, and several international delegations studied the implementation to replicate it in their own jurisdictions.

7.6 Case Study 6: Global Retailer's Multi-region Azure Architecture for High Availability and Edge Acceleration

The Digital Shelf Is the Storefront: Why Retail Can't Tolerate Downtime

In global retail, especially in ecommerce applications, latency is directly correlated with cart abandonment rates, and downtime translates instantly into lost revenue and damaged brand trust. In this case study, we analyze the cloud architecture of a global apparel retailer that moved from a fragmented set of regional hosting providers to a unified **multi-region Azure environment**, with the goal of delivering **sub-second latency experiences across continents** and ensuring **zero downtime during major sale events** like Black Friday or Singles' Day.

The retailer, with customers across North America, Europe, and Asia-Pacific, had to address key challenges:

- Eliminate **single-region dependency** that risked outages due to natural disasters or maintenance.

- Serve personalized, localized content with **minimal round-trip latency**.

- **Scale elastically** during peak demand while maintaining consistent performance.

- Meet **regulatory constraints** in markets like Germany (data residency) and Japan (latency SLAs).

CHAPTER 7 REAL-WORLD CASE STUDIES AND FUTURE TRENDS

To achieve this, the organization adopted a hybrid of **Active-Active** and **Active-Passive** models using Azure Front Door, Azure Traffic Manager, and regionally paired AKS and App Services with intelligent failover logic.

Core Design Pillars: Resilient, Localized, and Observed

The architecture was guided by three core principles:

1. **Resiliency Through Redundancy**

   - Applications were deployed across **four primary Azure regions** (East US, West Europe, Southeast Asia, Australia East).

   - Each region was configured with full workload autonomy, including its own instance of **AKS**, **Azure SQL Database**, and **Azure Redis Cache**.

   - **Geo-replication** of databases used **Active Geo-Replication** and **Zone Redundant Configuration (ZRC)** for stateful services.

2. **Localized Performance**

   - **Azure Front Door Standard/Premium** handled global traffic routing using real-time performance probes and TLS offloading.

   - Dynamic content was served via AKS behind Azure Application Gateway; static assets used Azure CDN endpoints tied to regional blob storage.

   - Product recommendations were cached at edge locations using **Azure Cache for Redis Enterprise with geo-distribution.**

3. **Comprehensive Observability**

   - **Azure Monitor**, **Application Insights**, and **Log Analytics** collected telemetry from every region.

   - **Distributed Tracing** was enabled across regions using OpenTelemetry and Azure Monitor Workspaces.

   - SLIs, SLOs, and Error Budgets were tracked via **Grafana Dashboards** connected to Azure Data Explorer.

CHAPTER 7   REAL-WORLD CASE STUDIES AND FUTURE TRENDS

Architecture Blueprint: Global Load Balancing with Active-Active AKS Clusters

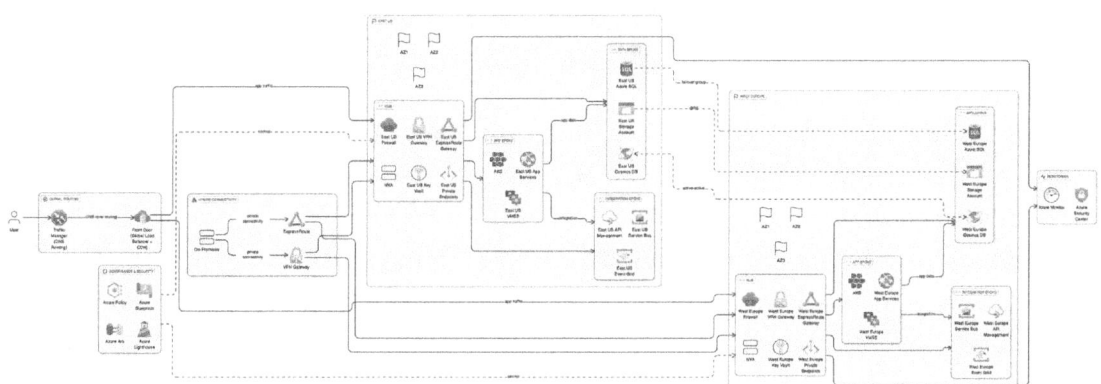

Figure 7-8. *Global Topology*

Figure 7-8 illustrates a globally distributed, production-grade Azure architecture designed to meet the stringent requirements of enterprise-scale workloads across multiple continents. At the heart of this topology is a multi-region deployment pattern spanning the **East US** and **West Europe**, each implementing a **hub-and-spoke virtual network model**. The **hub networks** in each region house critical infrastructure services such as **Azure Firewall**, **ExpressRoute Gateways**, and **Network Virtual Appliances (NVAs)**, establishing secure hybrid connectivity back to on-premises environments via VPN and private circuits.

Each **spoke network** segregates application workloads, data services, and integration components into distinct subnets, ensuring **network isolation and layered defense**. Workloads such as **Azure Kubernetes Service (AKS)**, **App Services**, and **Virtual Machine Scale Sets (VMSS)** run within the application tier, while the data tier hosts **Cosmos DB**, **Azure SQL**, and **Geo-redundant Storage Accounts**, all configured with **multi-region replication** and **automatic failover mechanisms**. Azure API Management (APIM), **Service Bus**, and **Event Grid** facilitate integration and asynchronous communication between services.

At the global control plane, **Azure Traffic Manager** performs **DNS-level routing** across regions, using **geographic and performance-based rules** to direct client requests to the most optimal regional deployment. In conjunction, **Azure Front Door** provides a globally distributed edge presence, accelerating content delivery while enforcing SSL termination and WAF policies.

237

Cross-region data synchronization is visualized through dedicated replication paths between database instances and storage services, supporting both **active-active** and **active-passive** modes based on workload needs. **Azure Key Vaults** are deployed regionally with private endpoints to ensure secrets never traverse public networks, while **Azure Policy**, **Azure Blueprints**, and **Azure Monitor** enforce governance and observability across all tiers.

The architecture also integrates **Azure Arc** and **Azure Lighthouse**, enabling centralized policy enforcement, logging, and identity management for distributed workloads irrespective of their physical location. This global topology exemplifies how enterprises can architect for **high availability**, **low latency**, **resilience**, and **compliance** in a cloud-native manner using the full breadth of Azure's hybrid and multi-region capabilities.

- **Azure Front Door** acts as the single entry point, performing SSL termination and routing based on latency, health, or geography.
- Behind Front Door, a nested **Azure Traffic Manager** handles region-level failover:
  - If East US is slow or unhealthy, traffic routes to Europe or Asia.
- Each region hosts
  - **AKS Cluster** with microservices for catalog, cart, payment, and user sessions
  - **Azure SQL Database Hyperscale** with Geo-Replication
  - **Azure Blob Storage with RA-GRS** for media content
- AKS uses **pod anti-affinity rules** and **zone-aware node pools** to protect against zone-level failures.

Infrastructure was provisioned using Terraform modules stored in a central GitOps registry. Region-specific parameter files controlled deployment scope, and deployments were rolled out via Argo CD.

Canary and Blue-Green Deployments at Scale

The retailer adopted progressive delivery techniques to reduce risk:

- **Canary Releases** were done using Azure Front Door routing rules with **traffic weights**.
- **Blue-Green Environments** were maintained in parallel for every region.
  - Blue: Current production
  - Green: Candidate version under test
- Feature flags were controlled centrally via **Azure App Configuration** and toggled in real time using Azure Functions.

In addition to rollback automation, all deployment workflows included **pre-deployment smoke tests**, **automated rollback on SLI violation**, and **Slack notifications** to global SRE channels.

Edge Acceleration and Personalization

Performance tuning extended to the edge:

- Azure Front Door Rules Engine injected user personalization headers (locale, device type) into upstream requests.
- AI-powered product recommendation engines were trained in **Azure Machine Learning** and deployed via ONNX models into **Azure Functions Premium**, replicated globally.
- Azure CDN was extended with custom rules to serve personalized images (e.g., gender-based banners, local currencies) using Azure Blob versioning and caching headers.

Latency benchmarks revealed

- 180ms global median page load time (TTFB), a 37% improvement over legacy setup
- Sub-1s cart render time for 85% of global users

Security and Compliance Across Borders

Security was enforced at every layer:

- **Azure WAF** with managed rules protected all ingress endpoints.
- **DDoS Protection Standard** was enabled per region.
- Regulatory compliance was mapped using Azure Policy and Azure Blueprints:
    - **GDPR (Europe)**: Regional data stores, anonymization, consent logs
    - **CCPA (California)**: Per-user deletion workflows via Logic Apps
    - **APPI (Japan)**: Latency SLAs and data residency constraints

Azure Security Center (now Microsoft Defender for Cloud) continuously scanned the environment and enforced recommendations.

Results: Business Impact and Operational Lessons

- **99.99% application availability** across all regions, even during two separate regional service incidents.
- **$22M revenue increase** attributed directly to improved conversion rates and customer retention from faster load times and regional reliability.
- **35% cost optimization** via Reserved Instances, autoscaling, and smart regional failover.

Lessons:

- Nested Traffic Manager profiles helped isolate regional outages without affecting global availability.
- Regional observability must be prioritized equally blind spots in Asia cost them five hours of missed alerts during an early outage.
- Active-Active requires cultural and procedural shift, not just technical change.

7.7 Case Study 7: Cloud-Native Banking—High-Frequency Trading (HFT) on Azure Confidential Compute

Speed, Security, and Secrecy: The Mandate of Modern Trading Platforms

In the realm of high-frequency trading (HFT), milliseconds are an eternity, and data is gold. This case study dives into the architecture of a leading investment bank that migrated a portion of its algorithmic trading workloads to Azure, leveraging **Confidential Compute**, **AKS with DPDK-accelerated nodes**, and **ultra-low latency networking**.

The bank faced three critical imperatives:

- **Execution Speed**: Orders must be executed faster than competitors. Every packet, every interrupt, every syscall counts.

- **Data Confidentiality**: Trading algorithms are crown jewels. Unauthorized access, whether internal or external, is unacceptable.

- **Deterministic Latency**: Even under load or noisy neighbor conditions, the system must perform consistently within tight bounds.

Azure was selected not just as a scalable cloud provider, but as an innovation partner offering hardened enclaves, programmable FPGA support, and industry-aligned compliance.

Security Beyond Encryption: Shielded Logic in the Cloud

Security was treated as first-class, not just encryption in transit or at rest but *in use*. The confidential computing model ensured:

- **Trading algorithms were loaded into SGX enclaves**, decrypted at runtime, and never persisted on disk.

- **Secrets like API keys, exchange credentials, and ML models** were injected securely via **Azure Key Vault with Managed HSM** and **Enclave Attestation Services**.

CHAPTER 7 REAL-WORLD CASE STUDIES AND FUTURE TRENDS

- **No debugger access, no remote login**, and strict RBAC policies enforced via **Azure AD Conditional Access**.

Compliance mappings included

- **SEC Reg SCI**, **FINRA audit trails**, and **ISO 27001**.
- Custom **logic in Azure Policy** ensured no drift in VM SKU or image integrity.

All enclave health telemetry was piped into a dedicated **Azure Monitor Workgroup** visible only to the bank's Security Operations Center (SOC).

Data Flow: From Market Feed to Order Execution in 15ms

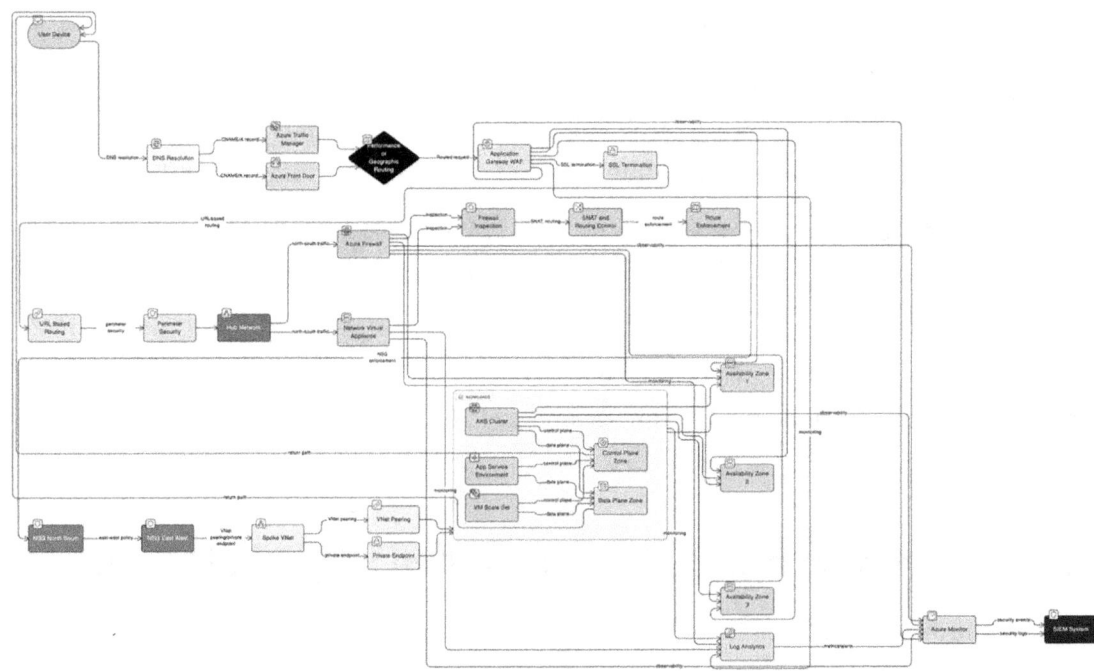

Figure 7-9. Packet-Level Flow

Figure 7-9 provides a detailed view of the **packet-level flow across the hybrid Azure enterprise architecture**, mapping the precise ingress, routing, inspection, and egress paths that a data packet traverses from user to workload. This visualization is critical for architects and security engineers who must validate how traffic is inspected, authorized, and routed across zones, networks, and enforcement layers.

The flow begins at the user's edge device, where DNS resolution is performed via Azure Traffic Manager or Azure Front Door, selecting the optimal region or endpoint based on performance, geography, or failover policy. From there, the packet enters Azure through the perimeter layer, typically an **Application Gateway with Web Application Firewall (WAF),** which handles SSL termination, path-based routing, and initial inspection.

Next, traffic passes through a **Network Virtual Appliance (NVA)** or **Azure Firewall** in the hub virtual network, where advanced policies such as threat intelligence filtering, outbound SNAT rules, and UDR-based routing are applied. The packet is then routed through peered VNets or private link endpoints to reach the workload tier, often an **AKS node pool, App Service Environment, or Virtual Machine Scale Set** residing in a spoke virtual network.

Importantly, Figure 7-9 distinguishes between **control-plane** traffic (e.g., from Azure DevOps or GitHub Actions) and **data-plane** traffic (e.g., user payload), showing how they follow separate paths and enforcement boundaries. It also visualizes lateral movement prevention mechanisms, including NSGs, service tags, and microsegmentation using Azure Policy or Defender for Cloud.

The diagram concludes with monitoring flows routed to **Log Analytics**, **Azure Monitor**, and **SIEM connectors**, ensuring observability and compliance across all tiers of the packet journey.

1. **Market Feed Ingestion**: Data from NYSE and NASDAQ co-lo facilities streamed via **Azure Event Hubs with Kafka protocol**.

2. **Preprocessing Layer**: Stateless pods normalize feed data and apply real-time filters.

3. **Signal Generator**: ML model running in an enclave container evaluates trade signals.

4. **Order Gateway**: Writes order requests to Azure Cosmos DB with low-latency indexing enabled.

5. **Execution Engine**: Posts the order to the exchange API and records the outcome.

Round-trip latency benchmarks:

- Market Feed ➤ Signal ➤ Order Execution: **13.7ms median**
- Failover to secondary region: **<2 seconds cold switch**

Observability in a Black Box World

While SGX enclaves restrict introspection by design, observability was engineered in layers:

- **Nonsensitive metrics** exposed via **Prometheus node exporters** on the host
- **Application health**, **SLAs**, and **latency histograms** exported via **OpenTelemetry sidecars**
- **Encrypted logging** routed to **Azure Data Explorer** and visualized in **Grafana**

To ensure reliability:

- **Synthetic transaction injection** simulated market events every minute
- **Chaos testing via Azure Chaos Studio** introduced network throttling and node failures

This helped validate that enclaves remained deterministic and latency budgets were preserved even under disruption.

Failover Strategy and Circuit Breakers

The architecture supported an Active-Passive regional topology:

- **Primary Region** (East US 2) hosted live trading
- **Secondary Region** (Central US) received real-time replication via Azure Cache for Redis and Cosmos DB multi-master
- **Nested Azure Traffic Manager profiles** controlled failover with near-zero RTO

A **circuit breaker layer** enforced

- Freeze trading if signal deviation crosses 3σ, where σ represents the standard deviation of a signal from its historical mean. Under a normal distribution, approximately 99.7% of observed values fall within three standard deviations of the mean σ; exceeding this threshold usually indicates an abnormal or potentially risky condition, triggering a fail-safe such as halting trades.

- Automatic rollback to the last known good signal model if the container crashes or fails attestation

- Email/SMS/Slack integration for trader alerting

Strategic Outcomes and the Path Ahead

- **Confidential compute unlocked new cloud use cases** once considered off-limits.

- Regulatory audits passed without exceptions unprecedented for a cloud-hosted HFT platform.

- AI models for predictive trade signals could now run securely on GPU-backed containers with **NVIDIA H100 Enclaves (preview)**.

What changed:

- From co-located, expensive, hard-to-scale metal ➤ to cloud-native, secure, scalable clusters

- From humans monitoring ➤ to ML watching ML

7.8 Case Study 8: Sustainability and GreenOps in Azure Enterprise Deployments

The Carbon Mandate: From Reporting to Architectural Responsibility

A Fortune 100 global pharmaceutical company, with over 120 Azure subscriptions spread across research, clinical, and commercial divisions, was mandated by the board to reduce IT-generated emissions by 35% in three years. This wasn't just about carbon reporting; it was a full-stack transformation in how cloud infrastructure was designed, provisioned, and operated.

GreenOps, the discipline of applying sustainability principles to cloud operations, emerged as the guiding approach. The company partnered with Microsoft's sustainability engineering group to track, visualize, and optimize its Azure resource footprint.

Key questions that shaped the initiative:

- Which Azure regions use the cleanest energy mix?
- How can we enforce carbon-efficient resource types through policy?
- What's the energy and emission cost per app service, per container, per GB?

Their objective was to build carbon-aware architectures without disrupting DevOps agility or regulatory compliance, especially critical in a regulated pharmaceutical context (GxP, HIPAA, ISO 13485).

Architecture Patterns for Sustainability

A new class of architectural blueprints emerged, optimized not just for scalability or HA but also **carbon efficiency**:

- **Region Selection Based on Carbon Intensity**: Using Azure Sustainability APIs, workloads were tagged for deployment in regions with the lowest marginal emissions (e.g., North Europe over East US).

- **Workload Right-Sizing Pipelines**: Every provisioning request, whether from Terraform, Bicep, or Azure Dev Portal, was intercepted by an Azure Function acting as a **Carbon Admission Controller**. It performed:

    - Instance right-sizing (e.g., Standard_D4s_v3 ➤ B2ms)

    - Scheduled de-provisioning of idle test environments

    - Optimization suggestion back to developers via Teams integration

- **Carbon Scoring on CI/CD Pipelines**

    - Build artifacts were tagged with estimated emissions based on compute time and region.

    - Application teams received a Green Score as part of their Azure DevOps pipeline dashboard.

- **Serverless-First Architecture**

    - Whenever possible, stateless workloads were deployed to **Azure Functions** or **Container Apps** using **Consumption Plans**, which scale to zero.

    - Long-running jobs were migrated to **Durable Functions**, breaking work into short, billable bursts.

CHAPTER 7 REAL-WORLD CASE STUDIES AND FUTURE TRENDS

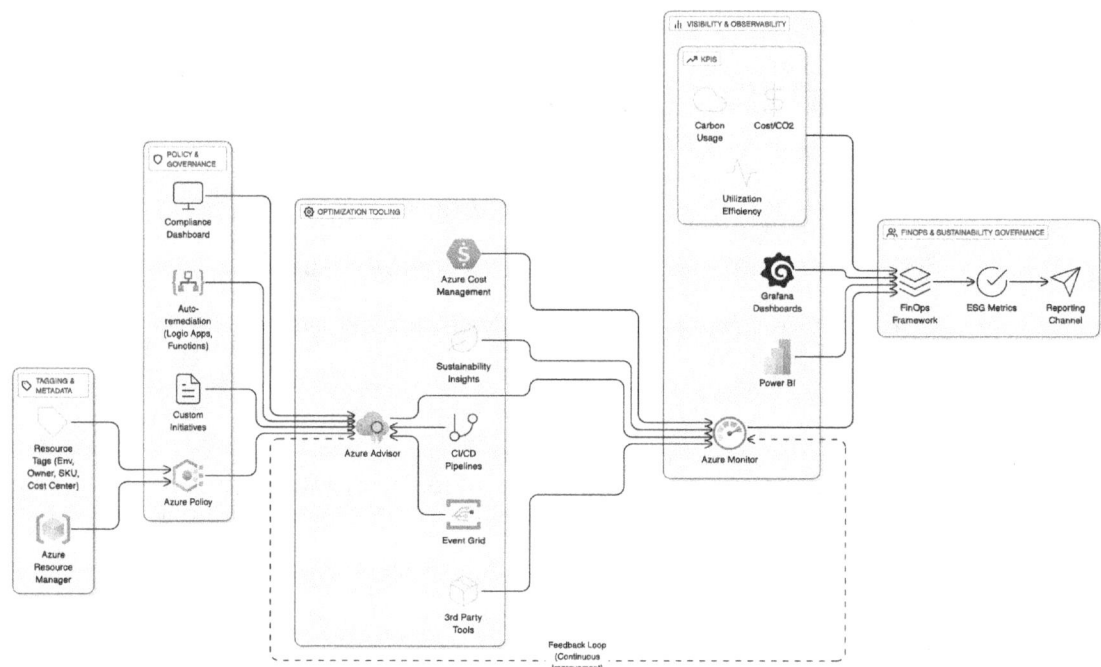

Figure 7-10. *"GreenOps Reference Architecture," Combining Proactive Tagging, Policy Enforcement, Optimization Tooling, and Visibility Layers*

Figure 7-10 illustrates the **GreenOps Reference Architecture**, a holistic framework designed to drive sustainable cloud operations by embedding environmental consciousness into every layer of the Azure ecosystem. This architecture integrates cost optimization, carbon awareness, and operational efficiency through a combination of tagging strategies, policy enforcement, automation tooling, and multilevel observability.

At the foundation of the diagram lies **resource tagging and classification**, which assigns metadata such as environment type, application ownership, energy efficiency ratings, and cost centers. These tags feed into **Azure Policy definitions** and **custom initiatives** that enforce guardrails, ensuring that noncompliant resources (e.g., underutilized VMs, over-provisioned SKUs, or non-Green SKUs) are flagged or auto-remediated.

The **optimization tooling layer** shows the integration of native services like **Azure Advisor**, **Cost Management + Billing**, and **Sustainability Insights**, alongside third-party tools that provide actionable recommendations. These tools are triggered on deployment pipelines, scheduled jobs, or via Event Grid/Logic Apps for real-time optimization loops.

The upper layers depict **governance and visibility**, including dashboards built on **Azure Monitor**, **Power BI**, or **Grafana**, visualizing KPIs such as energy usage, carbon footprint, and cost per application. Annotations highlight integration with **FinOps practices**, where GreenOps becomes a subset focused on aligning financial accountability with environmental sustainability.

Together, this architecture promotes an operating model where sustainability metrics are as first-class as performance or security, enabling enterprises to proactively manage not only their cloud cost, but their environmental impact, ensuring long-term operational and ecological viability.

Instrumentation and Visualization

Transparency drove behavioral change. The IT Sustainability Team implemented a carbon telemetry layer that collected, correlated, and visualized emissions per workload, team, and region.

Key tools and integrations included

- **Microsoft Sustainability Manager (MSM)**: Aggregated emissions data at scope 1, 2, and 3 levels

- **Azure Monitor Custom Metrics**: Added carbon metrics (e.g., grams CO_2e/hr) to standard dashboards

- **Grafana dashboards with a plugin for carbon scoring** per AKS workload, showing
    - Real-time consumption in CO_2e
    - Historical trends
    - App performance vs. energy efficiency index

All visualizations were integrated into **Executive Green Scorecards**, informing quarterly business reviews and R&D prioritization.

Policy As Guardrail: Carbon-Aware Governance at Scale

Governance was enforced through a layered **Azure Policy** approach:

- **Allowed Resource Types**: Teams could only provision VMs or services from an approved list of low-emission SKUs (e.g., D, B, E series over F or H).

- **Region Allow/Deny**: Based on live regional carbon data, noncompliant regions were denied unless justified via the exemption workflow.

- **Resource Expiry Policies**: Test environments tagged with env:dev were automatically scheduled for deletion after five days of inactivity.

These policies were nested under **Management Groups** corresponding to each business unit. Exceptions were granted through Azure Policy Exemption workflows, logged, and reviewed monthly.

Organizational Change: Embedding GreenOps into DevOps DNA

Beyond tooling, the hardest part was cultural. Developers were not accustomed to thinking about carbon footprints. The company addressed this through

- **Developer education sessions** on energy modeling and carbon metrics

- **Green sprint awards** for teams achieving a measurable reduction in emissions

- **Embedded sustainability advocates** in platform and infra teams

A centralized **GreenOps Guild** was formed, analogous to SRE or DevSecOps teams that drove practice evolution, published green design patterns, and reviewed architectural decisions through a carbon lens.

Outcomes and Metrics

In 18 months, the pharmaceutical company achieved:

- **41% reduction** in compute-related emissions across Azure
- **30% reduction** in non-production resource usage via automation
- **90% of new workloads deployed in carbon-efficient regions**

They were able to align sustainability with compliance and performance:

- Clinical data remained in sovereign, low-carbon regions.
- App performance was unaffected (in some cases improved due to optimization).
- Auditable carbon data was available for internal and ESG reporting.

And most critically, they transformed cloud cost into a *value* discussion not just in dollars but in decarbonization.

7.9 Edge + AI at Scale: Intelligent Applications with Azure Percept and Azure Arc

Introduction: Intelligence Where Data Is Born

In traditional enterprise architectures, intelligence lived in centralized data centers or in the cloud, far removed from the real-world environments where events were happening. Whether it was machinery emitting vibration signals on a factory floor, traffic cameras capturing congestion patterns, or retail shelves needing restocking, the prevailing design assumed that all data must flow to the cloud before decisions could be made.

This architecture created latency, reliability, and cost concerns, especially for high-frequency data or time-sensitive decisions. In contrast, edge computing brings compute, analytics, and machine learning inference closer to the source of data. When combined with AI, it opens up a new frontier of intelligent, autonomous applications that operate in real time without a roundtrip to the cloud.

Azure enables this at scale through a convergence of services: **Azure Percept** for edge AI hardware and vision, **Azure Arc** for hybrid and edge Kubernetes management, and **Azure Machine Learning (AML)** for life cycle management of models trained in the cloud but deployed to the edge.

This section walks through a real-world case study: a multinational manufacturing company using **Azure Edge + AI** to detect faults in assembly lines across global factories, reduce unplanned downtime, and automate quality assurance using AI vision.

Enterprise Use Case: Predictive Quality Control in Global Smart Factories

The company's challenge was twofold:

- **Ensure consistent product quality across dozens of plants in Asia, Europe, and North America.**
- **Avoid expensive unplanned downtime caused by equipment faults or undetected defects.**

While existing monitoring solutions collected metrics (e.g., temperature, vibration), they were reactive. Human inspectors reviewed images and audio recordings manually. This led to

- Inconsistent results due to human fatigue or bias
- Delays between defect occurrence and correction
- High operational costs of inspection teams

The goal was to implement a **real-time, intelligent quality control system**, using

- Cameras for image recognition of product defects
- Microphones for acoustic anomaly detection
- Vibration sensors for equipment degradation detection

Each data point would be processed at the edge with minimal latency. AI models trained in the cloud would run directly on factory-floor devices even in areas with intermittent connectivity.

CHAPTER 7   REAL-WORLD CASE STUDIES AND FUTURE TRENDS

Architectural Blueprint: Azure Edge AI with Arc and Percept

The architecture was designed with three tiers.

1. Perception Layer (Azure Percept DK + Custom Edge Devices)

- **Azure Percept DK**: Mounted on manufacturing lines to detect visual anomalies (e.g., scratches, misalignments).

- **Industrial Raspberry Pi/NVIDIA Jetson devices**: Process real-time vibration and audio data for anomaly detection.

- These devices ran containers deployed via **Azure IoT Edge runtime** and executed inference using ONNX models.

2. Edge Orchestration Layer (Azure Arc + AKS Edge Essentials)

- **Arc-Enabled Kubernetes Clusters**: Deployed in regional on-prem data centers near factories.

- These clusters received models from Azure Machine Learning and workloads from GitHub Actions CI/CD pipelines.

- Monitoring agents collected telemetry via **Azure Monitor for containers** and reported edge health to central teams.

3. Cloud Layer (Azure ML + Data Lake + Synapse)

- **Model training and retraining** occurred in Azure ML, using datasets uploaded by edge nodes.

- A central **Data Lake Gen2** hosted structured and unstructured data, and **Synapse** enabled cross-factory analytics.

- **Power BI dashboards** displayed live defect heatmaps, sensor anomalies, and model accuracy metrics.

Model Life Cycle and Deployment with Azure ML and IoT Edge

Edge AI is only effective when the model life cycle is robust, repeatable, and governed. The company adopted the following pipeline:

- **Training**: Cloud-scale training using historical defect datasets in AML.

- **Validation**: Human-in-the-loop validation with labeling tools and active learning feedback.

- **Packaging**: Models exported to ONNX, wrapped in Docker containers with inference code.

- **Distribution**: GitHub Actions published containers to **Azure Container Registry (ACR)**.

- **Deployment**: ACR images pulled into Arc-managed AKS clusters and IoT Edge devices.

The use of **GitOps with Flux and Arc** enabled declarative, policy-enforced deployments to hundreds of distributed edge locations, with rollback support and tamper detection.

Security, Reliability, and Offline Operation

To meet manufacturing-grade standards, the following measures were enforced:

- **Offline Resilience**: Devices cached models and queue events for batch upload when the cloud was unreachable.

- **Tamper Protection**: Azure Defender for IoT was enabled to detect unauthorized firmware or container changes.

- **Data Encryption**: Sensor data was encrypted on disk (using TPM modules) and in transit (TLS 1.2).

- **Model Provenance**: MLFlow-based lineage tracking ensured each deployed model was traceable to training code, dataset, and commit ID.

In environments with Zero Trust constraints, edge devices used **certificate-based authentication** and reported to **Azure Security Center** for risk posture evaluation.

Results and Measurable Impact

The transformation delivered massive improvements:

- **Defect detection time reduced from 45 minutes to 3 seconds.**
- **Unplanned downtime decreased by 36% year-over-year.**
- **Human inspection costs lowered by 50%, freeing resources for strategic QA.**
- **Cross-factory model tuning** became real-time, enabling collaborative model evolution across geographies.

Moreover, site engineers embraced the new system due to its intuitive dashboards and AI-assisted diagnostics.

7.10 Lessons Learned: Common Anti-patterns in Azure Transformations

Introduction: The Cost of Missteps in Cloud Journeys

Enterprise cloud adoption, particularly in regulated or large-scale environments, is a high-stakes transformation. The promise of elasticity, scalability, and innovation is often undermined by deeply ingrained legacy practices, organizational silos, or rushed execution. In Azure environments, this manifests as architectural drift, governance breakdowns, cost overruns, and security exposure, none of which are caused by Azure itself but by how Azure is used.

This section consolidates hard-earned lessons from dozens of Azure engagements across verticals. It outlines real-world anti-patterns, not theoretical flaws but systemic issues observed in practice that compromised availability, security, performance, and maintainability. For each anti-pattern, we examine the root causes, its architectural symptoms, and mitigation strategies grounded in Azure-native best practices.

Anti-pattern 1: Lift-and-Shift Without Re-platforming

Symptom

Workloads are migrated as-is from on-premises into Azure VMs without any reconfiguration. Legacy dependencies (e.g., Windows Server 2008, hard-coded file shares, or SQL Server with unsupported versions) are retained.

Consequences

- Poor performance due to a mismatch with Azure VM types or disk tiers
- High costs compared to PaaS alternatives
- Limited observability and automation capabilities

Remediation

- Implement re-platforming to **Azure App Services, Azure SQL, or Azure Kubernetes Service**.
- Use **Azure Migrate** with dependency mapping to inform modernization.
- Adopt the **CAF migration stages**: Assess ➤ Migrate ➤ Optimize ➤ Secure ➤ Govern.

Anti-pattern 2: Subscription Sprawl Without Management Group Strategy

Symptom

Each application team creates its own subscriptions ad hoc. There's no global policy enforcement, naming standard, or consistent tagging.

Consequences

- Governance complexity increases over time.
- Security baselines (e.g., Defender for Cloud) vary across environments.
- Billing lacks visibility, making showback/chargeback models infeasible.

Remediation

- Implement a **Management Group hierarchy** aligned with business units, environments, and geographies.
- Enforce **Azure Policy** and **blueprint artifacts** at the Management Group level.
- Use **Azure Cost Management + tags** for spend attribution.

Figure 7-10 shows a common misaligned subscription sprawl before and after alignment under a centralized management group hierarchy.

Anti-pattern 3: Overreliance on Portals, Underinvestment in IaC

Symptom

Most infrastructure is manually provisioned via the Azure Portal. Scripts exist but are inconsistent or non-repeatable. There is little to no versioning or Git-based review.

Consequences

- Environments drift over time due to manual changes.
- Recovery is hard in DR or test environments.
- Onboarding new environments becomes slow and error-prone.

Remediation

- Use **Terraform or Bicep** to declare infrastructure.
- Standardize modules, and enforce pull requests via **GitHub Actions or Azure DevOps Pipelines**.
- Integrate **policy as code** to guardrail deployments.

Anti-pattern 4: Ignoring Azure Regions and Availability Zones Strategy

Symptom

All resources deployed into a single region without considering zone redundancy or regional failover. No secondary region is defined for DR.

Consequences

- Service outages in that region impact all workloads.
- Data sovereignty or compliance policies may be violated.
- Latency-sensitive workloads underperform in distributed user bases.

Remediation

- Adopt **Active-Active or Active-Passive multi-region topologies**.
- Use **Zone-redundant services** (e.g., Zone-redundant storage, SQL Managed Instance).
- Architect with **Traffic Manager, Azure Front Door, and paired regions**.

Anti-pattern 5: Skipping Identity and RBAC Foundations

Symptom

Azure AD is used minimally. Contributors or Owners are assigned to individuals instead of groups. External users may have access without conditional access policies.

Consequences

- Elevated risk of privilege escalation or insider threat.
- Auditing becomes opaque due to a lack of centralized control.
- CI/CD pipelines may leak secrets due to misconfigured identity roles.

Remediation

- Use **Azure AD Security Groups and Privileged Identity Management (PIM)**.
- Adopt **role-based access control (RBAC)** with least-privilege design.
- Implement **Managed Identities** for service-to-service authentication.

Anti-pattern 6: Neglecting Observability and Telemetry

Symptom

Applications are deployed without consistent use of **Application Insights**, **Log Analytics**, or **Azure Monitor**. Errors go unnoticed until customer complaints surface.

Consequences

- MTTR (Mean Time to Resolution) is high.
- No proactive detection of performance degradation.
- SLAs and compliance objectives are breached due to a lack of visibility.

Remediation

- Enable **end-to-end observability** using Azure Monitor, App Insights, and distributed tracing.
- Route logs to **Azure Data Explorer or Sentinel** for long-term storage and analytics.
- Set up **alert rules and action groups** for auto-remediation.

Anti-pattern 7: Over-isolation by Region and Network Segmentation

Symptom

Cloud architects enforce strict region-based and network-based isolation without a cohesive integration strategy. Each region operates as a silo with its own VNet, policies, identity setup, and resources. Inter-region or inter-environment communication is discouraged or disabled entirely, even for shared services or governance tools.

Consequences

- Operational fragmentation across environments, leading to duplicated services and configurations
- Increased costs due to redundant infrastructure like multiple Key Vaults, Log Analytics workspaces, and Azure Policy definitions
- Lack of centralized governance and visibility, especially in security operations and incident response
- Hard-to-scale deployments and friction for platform engineering teams who must maintain multiple baselines

Remediation

- Consolidate shared services such as Azure Monitor, Azure AD, and management tooling using a centralized or federated architecture.
- Use **Azure Virtual Network Manager (AVNM)** to enforce global security rules while allowing controlled connectivity between regions.
- Adopt a hub-and-spoke network model with **regional spokes** and a shared **global control plane** (e.g., Azure Lighthouse, Azure Policy, Defender for Cloud).
- Balance isolation with integration by defining **clear network boundaries** but maintaining observability and operational consistency through shared pipelines, identity, and monitoring.

CHAPTER 7   REAL-WORLD CASE STUDIES AND FUTURE TRENDS

Note   While isolation is a valid pattern for tenant separation, compliance zones, or blast-radius control, excessive or uncoordinated isolation becomes an anti-pattern that undermines the very agility and scale the cloud promises.

Closing Reflections: The Discipline of Cloud Maturity

None of the above anti-patterns is merely a technical oversight; they are organizational indicators. They reflect the maturity of cloud operating models, the rigor of architecture review boards, and the culture of continuous improvement. Azure, as a platform, offers the tools to mitigate these issues, but it is the responsibility of enterprise architects and cloud governance leads to embed those tools in practice.

Architecture in Azure is not a static diagram; it is a living contract between business goals and technological execution. Recognizing anti-patterns early and remediating them decisively ensures that the cloud becomes an accelerator, not an obstacle.

7.11 Future Trends in Azure Infrastructure and Architecture

Introduction: Evolving Beyond Infrastructure

As the global digital economy grows increasingly interdependent and AI-driven, the expectations from cloud platforms like Azure are rapidly shifting. It is no longer sufficient to be merely cloud-enabled; organizations are expected to be cloud-native, cloud-smart, and cloud-resilient. The underlying infrastructure is becoming invisible yet indispensable, abstracted through platforms, platforms-as-a-service, and increasingly autonomous orchestration layers.

This section explores the leading-edge trends shaping Azure's future landscape. These aren't abstract forecasts; they are already being piloted in forward-looking enterprises and quietly altering how cloud infrastructure is planned, deployed, and governed. The discussion spans intelligent infrastructure, AI integration, sustainability, confidential computing, and the role of edge and spatial computing. Each trend represents a design inflection point that architects and decision-makers must actively embrace or risk being outpaced by disruption.

AI-Native Infrastructure

With the rise of large language models (LLMs), generative AI, and machine learning at scale, infrastructure is becoming AI-aware. Azure is not merely hosting AI models; it is optimizing entire infrastructure pipelines for them.

Key Evolutions

- **Azure AI Studio** and **OpenAI on Azure** integrate AI pipelines with infrastructure provisioning.
- **GPU-aware autoscaling** in AKS, with NVIDIA A100/H100 support.
- **Intelligent caching and tiering** for AI data pipelines using Azure Data Lake Gen2 and Azure NetApp Files.

Implications

Architects must now design for hybrid AI workloads, with infrastructure that dynamically scales compute, balances inference and training traffic, and manages cold/warm GPU pools with cost optimization.

Confidential Computing and Zero Trust Fabric

Security is no longer a perimeter concern; it is embedded into every byte of infrastructure. Azure is leading with confidential computing, enabling sensitive workloads to run in trusted execution environments (TEEs).

Technologies to Watch

- **Azure Confidential VMs** (based on AMD SEV-SNP and Intel SGX)
- **Always Encrypted with secure enclaves** for SQL and sensitive data
- **Confidential containers in AKS** using Kata containers and gVisor

Architect's Takeaway

Expect to see infrastructure blueprints requiring **workload attestation**, **hardware-backed trust**, and **infrastructure-as-policy** (IaP) models for critical workloads, especially in financial services and defense.

Sustainable and Carbon-Aware Architecture

Azure has committed to being carbon negative by 2030, and that commitment is now reflected in service architecture.

Enablers

- **Microsoft Sustainability Manager** with real-time emission insights.
- **Carbon-aware workload scheduling** APIs in Azure Batch and Azure Kubernetes Service (AKS).
- **Green-region deployment insights**, developers, and architects will soon choose regions not just for latency but for carbon intensity.

Architectural Practice Shift

Sustainability will be a **first-class nonfunctional requirement**. Design decisions from ephemeral vs. persistent compute to data replication and logging verbosity will be made with emissions and sustainability scores in mind.

Edge, 5G, and Spatial Computing

The future of infrastructure is not centralized; it's massively distributed. Azure Arc, Azure Private MEC, and Azure Percept are establishing Azure as an edge-native platform.

Emerging Patterns

- **Edge-native clusters** managed by Azure Arc for manufacturing and healthcare

- **5G-powered edge microservices** using Azure Private MEC and core packet services
- **Holographic and spatially aware apps** using Azure Spatial Anchors and Mixed Reality services

Implications

Architects must extend the definition of infrastructure to include edge devices, mesh networks, AR/VR endpoints, and low-latency physical-world integration. Azure's fabric becomes ubiquitous, stretching across data centers, IoT devices, and human interaction spaces.

GitOps, Platform Engineering, and Internal Developer Platforms (IDPs)

The CI/CD model is maturing into a **continuous everything** model. With Azure Kubernetes Service and GitHub Copilot, organizations are evolving toward platform teams that own paved paths for developers.

New Operational Constructs

- **Azure DevOps + GitHub Actions + Bicep/Terraform + FluxCD** for full GitOps pipelines
- **Azure Deployment Environments** with self-service developer sandboxes
- **Internal Developer Platforms (IDPs)** using Backstage, integrated with Azure services

Strategic Shift

DevOps is no longer a practice; it's a **product**. Architects must design infrastructure that serves both as a product delivery engine and a platform that improves the developer experience.

Serverless Infrastructure and Event-Driven Meshes

As business logic grows more reactive and asynchronous, serverless patterns and eventing backbones become the foundation of architecture.

Accelerating Constructs

- **Dapr sidecars in AKS** and Azure Container Apps
- **Event Grid Meshes** across regions and subscriptions
- **Functions-as-platform** with durable workflows and bindings to AI, databases, and external systems

Strategic Planning

Enterprise architecture will move toward **composable, event-centric microservices**, treating infrastructure as invisible glue. Observability, latency guarantees, and message durability will be foundational design pillars.

Conclusion: The Architect As Futurist

Future-proofing infrastructure is no longer about choosing the "most scalable" option; it is about designing systems that can evolve with uncertainty. Azure, as a platform, is growing not just in size but in *dimensionality,* spanning core, edge, AI, sustainability, and beyond.

The enterprise architect of tomorrow must balance technical foresight with pragmatic execution. The best cloud infrastructure will not be the most powerful or most expensive; it will be the one most aligned with the organization's purpose, adaptable in the face of unknowns, and capable of unlocking human potential through automation and intelligence.

7.12 Closing Reflections: Architecture Beyond the Horizon

In the evolving digital fabric of modern enterprises, Azure has emerged not only as a cloud provider but as a strategic enabler, a platform where infrastructure, security, development velocity, and innovation intersect. This chapter has journeyed through the depths of that transformation, from large-scale migrations and real-world governance practices to the future-facing paradigms of AI-native design, confidential computing, and edge-distributed intelligence.

Each case study has demonstrated a critical truth: enterprise architecture is not merely about technical correctness. It is a living discipline rooted in business alignment, operational readiness, organizational behavior, and strategic foresight. From deploying virtual networks at planetary scale to fine-tuning traffic flows with nested failover logic, architects must design with empathy for both systems and people. The cloud is not a destination; it is a medium for continual reinvention.

As we close this chapter, it is worth revisiting a foundational principle: **cloud architecture is not about predicting the future perfectly but about preparing for it resiliently**. The architectures that survive and thrive are not the most rigidly controlled but the most *evolvable*, governed by policy yet flexible in execution, secured by principle yet generous in enablement.

7.13 Summary

In this chapter, we explored the convergence of architectural principles and production realities through a series of real-world case studies and forward-looking strategies. As the Azure ecosystem continues to mature, enterprises are no longer experimenting with the cloud; they are refining it. This chapter reflected that maturity by showcasing applied patterns in sectors like financial services, healthcare, ecommerce, and global SaaS platforms, all of which demand production-grade resilience, compliance, observability, and cost governance.

Each case study examined a unique architectural concern, whether it was achieving sub-second RTOs using Azure Site Recovery, optimizing cloud spend with GreenOps practices, or deploying latency-sensitive workloads across multiple Azure regions with BGP and ExpressRoute integration. We highlighted how technical decisions are increasingly intertwined with business KPIs, regulatory frameworks, and global expansion strategies. These examples underscored that in real-world scenarios, perfect blueprints rarely exist; only adaptive patterns grounded in context, constraints, and continuous improvement.

The chapter also unpacked operational models and deployment topologies that transcend static architecture. Multi-region failover using active-active deployments, compliance scaffolding using Azure Blueprints, and the introduction of tools like Azure Chaos Studio and confidential computing capabilities revealed that enterprise-grade reliability now extends beyond uptime into verifiability, resilience testing, and data assurance.

Furthermore, we explored emerging architectural imperatives: observability as a discipline, cost optimization as a culture (GreenOps), and policy-as-code as a pillar of automated governance. With the rise of platform engineering, SRE disciplines, and decentralized DevOps models, Azure infrastructure must now be designed as a programmable, observable, and policy-enforced substrate, one that balances autonomy with control.

Index

A

Active-active multi-region topologies, 89
AI-native infrastructure, 262
Anti-patterns
    availability zone strategy, 258
    lift-and-shift without platforming, 256
    observability/telemetry, neglect, 259
    portals, 257, 258
    region-based and network-based isolation, 260
    skipping identity/RBAC foundations, 258
    subscriptions, 256, 257
Application Gateway, 95
Application Insights, 157, 163
Application Security Groups (ASGs), 87
Autoscaling, 122
Availability Sets, 90, 118
Availability Zones, 118
Azure Active Directory (AAD), 67, 126, 199, 201
Azure Advisor, 169, 170
Azure API Management (APIM), 237
Azure Application Gateway Ingress Controller (AGIC), 126
Azure App Service, 4
Azure Arc, 82, 252
Azure Container Apps (ACA), 4, 205
Azure Container Registry (ACR), 129, 234, 254
    CI/CD pipeline, 147
    cleanup/retention/policy enforcement, 151
    crictland Docker gRPC, VMSS runtime hygiene, 150
    enforcing base image governance, 148
    hub-and-spoke network topology, 152–155
    immutable tagging, 150
    robust container image life cycle, 147
    zero-day vulnerabilities, 152
Azure core services
    compute, 3, 4
    foundational services, 3
    networking, 6
    storage, 5
Azure Cosmos DB, 4
Azure Data Factory, 232
Azure Data Lake Storage Gen2, 233
Azure DevOps pipelines, 45
Azure Edge + AI
    architectural blueprint, 253
    enterprise use case, 252
    ML/IoT edge, 254
    results/measurable impact, 255
    security/reliability/offline operations, 254
Azure Event Grid, 205
Azure ExpressRoute, 74, 87
Azure Front Door, 111, 119
Azure Kubernetes Service (AKS), 4, 7, 49, 105, 119, 121, 159, 205, 218, 221, 263
    application performance tuning, 140
    CA, 137

INDEX

Azure Kubernetes Service (AKS) (*cont.*)
    CI/CD pipelines, 123, 142
    clusters, security
        control plane, 130
        multilayered and continuous, 128
        network security architecture, 128
        network segmentation/zero trust access, 133
        node plane, 130, 131
        real-world application, 133, 134
        secret credentials, 132
        traffic paths, 129
        workload level security policies, 132
    control plane *vs.* node plane, 124, 125
    core components, 122
    deployment architecture, 123
    deployment strategies, 144
    GitOps, 143
    HPA, 137, 139
    identity/role based access, 126
    microservices-based systems, 121
    namespaces, 123, 124
    networking and identity providers, 121
    node pools/OS options, 126
    node pool strategies/cost-aware scaling, 140
    real-time sports app, 141
    real-world application, 127, 146
    secret management/environment configuration, 145
    VPA, 139
    workload primitives, 125, 126
Azure Machine Learning (AML), 252
Azure Monitor, 157
Azure NetApp Files, 10
Azure policies, 14
    automated remediation workflows, 178, 179
    Azure governance architecture, 177
    DevOps/IaS, 183
    enforcing governance at scale, 180, 181
    Microsoft Defender, cloud, 182
    PCI-DSS compliance, 183
    remediation/compliance tracking, 181, 182
Azure Policy approach, 250
Azure Resource Manager (ARM) Template, 27
Azure Service Bus, 7, 205
Azure Service Group (SG), 11
Azure Site Recovery (ASR), 97, 98, 115, 116, 118, 190
Azure Stream Analytics (ASA), 227
Azure Table Storage, 4
Azure Traffic Manager, 112, 119
Azure Virtual Network Manager (AVNM), 9, 85, 87

B

Backup and data protection strategies
    AKS workload, 105, 106
    data integrity or restorability, 102
    governance/monitoring/compliance, 107, 108
    immutability/version control, 104
    PITR, 103, 104
    policy-based protection at scale, 103
Bicep, 27, 45, 116

C

Change control boards (CCBs), 38
Cloud Adoption Framework (CAF), 200
Cloud-native banking
    failover strategy/circuit breakers, 244, 245

INDEX

modern trading platforms, 241
observability, 244
security, 241, 242, 244
Cloud Center of Excellence (CCoE), 197
Cloud Native Computing Foundation (CNCF), 122
Cloud provider, 266
Cluster Autoscaler (CA), 137, 156
Compliance-by-design model, 233
Conditional Access Policies, 9
Confidential computing, 263
Consumption-based model, 167
Container runtime interface (CRI), 148
Continuous integration and continuous deployment (CI/CD) pipeline, 214
Continuous integration and delivery (CI/CD) pipelines, 37
Cost optimization, workloads
  advisor, 170, 171
  architecture, 168
  Azure cost management, 170
  cost efficiency, 172, 173
  cost governance, 175
  FinOps, 167
  foundational components, 168, 169
  multicloud strategies/FinOps, 176
  RIs/savings plans, 171
  spot VMs, 172
  storage/networking costs, 174, 175
Customer-managed keys (CMKs), 9, 37, 70, 116, 133, 189

D

Declarative model, 21
Defense-in-depth model, 55
Defense-in-Depth security architecture, 60, 80

DevSecOps, 122
Disaster recovery (DR), 118
  ASR
    failback/cost optimization, 101
    failover with recovery plans, 100
    localized failures, 98
    replication and recovery architecture, 98, 99
    security/policy integration, 101, 102
Distributed Denial of Service (DDoS) attacks, 56, 57
Domain-Driven Design (DDD) approach, 218
Dynamic site acceleration (DSA), 221

E

Ecommerce
  Azure architecture, 218, 221, 222
  business impact, 224
  IaC and deployment strategy, 223
  microservices, 217
  Peak Readiness, 223
  personalization, 222
Electronic Health Record (EHR) systems, 230
Enterprise deployments
  carbon-aware architectures, 246
  carbon-aware governance, 250
  GreenOps into DevOps DNA, 250
  instrumentation/visualization, 249
  sustainability, architecture patterns, 246–248
Enterprise-Scale Landing Zone (ESLZ) methodology, 200
Event Grid, 7, 237
Event Hubs, 232

INDEX

F

Fast Healthcare Interoperability Resources (FHIR), 230
Financial operations (FinOps), 193
Financial services
    business drivers/transformation goals, 196
    CI/CD and IaS, 202–204
    hub-and-spoke network model, 201, 202
    identity and management, 202
    landing zones, 196
    migration workloads, 204–206
    scalable landing zones, 197–200
    security/risk/compliance controls, 207
FinTech, 54
Firewall, 55, 56
Fully qualified domain names (FQDNs), 61, 80

G

Gatekeeper, 132
Geo-replicated data services, 119
GitOps, 38, 42, 45, 122, 143, 156, 264
Global retailer's multi-region Azure architecture
    business impact, 240
    canary/blue-green deployments, 239
    challenges, 235
    edge acceleration/personalization, 239
    global load balancing, 237, 238
    principles, 236
    security/compliance, 240
Governance and compliance, 14
    Azure policy, 14, 15
    cost management/guardrails, 16
    management groups/hierarchical control, 14
    RBAC, 15, 16
    regulated industries, 16
GreenOps, 246, 248

H

HashiCorp Configuration Language (HCL), 22
Healthcare
    compliance-by-design model, 233
    DevOps, 234
    ecosystems, 234, 235
    EHR, 230
    FHIR-first/API/Cloud-native, 230
    smart app integration, FHIR hub, 231–233
High availability (HA)
    architecture, 96, 97
    availability set, 90
    fault domains, 90
    standard load balancer, 95
    VMSS, 90, 91, 93, 94
High availability (HA) and disaster recovery (HA/DR), 87
    resilience, 89
High-frequency trading (HFT), 241
Horizontal Pod Autoscalers (HPA), 137, 156, 218
HSM-backed keys, 9
Hub-and-spoke virtual network architecture, 201
"Hub-network" module, 35
Hybrid cloud networking
    application/data layer, 81, 82
    Azure Arc, 82
    Azure ExpressRoute, 75, 76

decision matrix, 77, 78
DNS, 82
ExpressRoute, 74
internet-based secure connectivity, 75
monitoring layer, 81
optimal strategy, 78, 80
perimeter security layer, 80
shared services, 83
Virtual Network Gateway, Bicep, 83, 84
VPN Gateway, 73-75
Hybrid trust model, 209

I, J

Immutable container images, 156
Immutable tagging, 150
Infrastructure as a Service (IaaS), 3
Infrastructure as Code (IaC), 18, 117, 185, 199, 214, 234
   benefits, 20
      declarative *vs.* imperative model, 21
      Devops and CI/CD platforms, 22
   Bicep/ARM template
      automating deployments, 27, 28
      choosing right mode, 31, 32
      CI/CD integration/linting, 33
      complete mode, 30, 31
      deployment modes/template specs, 28, 29
      incremental mode, 29
      parameterization/secret management, 33
   CI/CD pipelines, 19
   cloud-native architecture, 19
   enterprise environments, 20
   GitOps, 42-44
   large scale environments, 34-38
   matrix strategy, 40

   multiple Azure regions, 39
   real-world use cases, 44
   SDP, 41, 42
Infrastructure-as-policy (IaP) models, 263
Intelligent quality control system, 252
Internal Developer Platforms (IDPs), 264
Internet Edge, 80
Internet Key Exchange (IKE), 73
Internet Protocol Security (IPsec), 73
IPsec/IKE VPN tunnels, 83

K

KEDA-based event triggers, 4
Kusto Query Language (KQL), 160, 162

L

Large language models (LLMs), 260
Log analytics/Azure monitoring, 157, 162
   App Insights, 160
   Application Insights, 163, 164
   containers extension, 159
   end-to-end observability architecture, 159
   metrics/alerts/visualization, 162, 163
   observability architecture, 161
   real-world application, 164-167
   workbooks, 158
   workspace, 162

M

Management group and landing zone strategy, 86
Microservices, 2, 4
Microsoft 365, 67

INDEX

Microsoft Azure
    cloud, 2
    cloud-native infrastructure, 1
    designing infrastructure
        enterprise resilience, 10
        scalability, 7, 8
        security, 9
    enterprise, logical constructs, 11–13
    enterprise use cases, 3
Microsoft Cloud Adoption Framework (CAF), 197
Monolithic systems, 1
Multi-factor Authentication (MFA), 9, 67, 71
Multi-region deployments
    active-active multi-region architecture, 111, 112
    active-passive *vs.* active-active deployment models, 109
    automation/recovery readiness, 116
    Azure Traffic Manager, 112, 113
    data synchronization/consistency models, 110
    design, 108
    global traffic distribution and routing, 110
    governance/compliance/residency, 116
    nested Azure Traffic Manager profiles, 113, 114

N

National health systems
    DevOps, 214–216
    governance/audit/compliance assurance, 216
    identity federation/pseudonymized data design, 212
    legacy systems *vs.* hybrid models, 213
    nation-sized microservices, 209–211
    observability/resilience engineering, 213
    time and constraints of policy, 208, 209
    transformation impact, 217
Networking and security
    AVNM, 85, 86
    FinTech, 54, 55
    private endpoints, 50–53
    routing tables, 47
    subnets, 49
    VNet, 48, 49
    Zero Trust principles, 47
Network Security Groups (NSGs), 6, 9, 49, 64, 80, 85, 87, 101, 129, 180
    application/data layer, 62–65, 67
    diagnostic logs, 58
    vs. firewall rules, 58
    government portal, 59
    micro-segmentation, 57
    monitoring layer, 61, 62
    networking layer, 61
    perimeter security layer, 61
    rules, 59
Network topology, 9
Network Virtual Appliances (NVAs), 6, 82, 237, 243
Node pools, 126

O

Open Policy Agent (OPA), 132, 203, 215

P, Q

Path-based routing, 97
Platform-as-a-service (PaaS), 4, 62, 81, 205

Pod Security Admission (PSA) policies, 132
Point-in-time restores (PITR), 118
Policy-as-code and posture-as-a-service model, 177
Policy-as-code and secure-by-default approach, 179
Policy-based deployments, 117
Policy-driven strategy, 102
Policy sets, 178
PostgreSQL, 103
Private Endpoints, 129
Privileged Identity Management (PIM), 9, 16, 199, 201, 212, 228
Protected Health Information (PHI), 234

R

Real-time risk engine, global trading
  continuous deployment, 229
  legacy bottlenecks/Azure motivation, 225
  performance engineering, 228
  regulatory confidence, 229
  risk at millisecond speed, 224
  streaming-based risk engine, 225–227
  zero trust/high compliance, 227
Recovery Point Objective (RPO), 98, 99
Recovery Time Objective (RTO), 98
Regulatory compliance, industries
  auditing and evidence collection, 186
  Azure Blueprints, 185
  certifications, 188
  compliance tools, 190
  control layers, 187
  design considerations, 188–190

DINE policy, 191
frameworks, 184
healthcare, 192
IaC, 193
PCI DSS and HIPAA, 185
policy enforcement, 186
Reserved Instances (RIs), 171
RESTful APIs, 4
Role-based access control (RBAC), 12, 14, 15, 17, 67, 68, 101, 107, 126, 130, 185, 189, 197, 233
Runbook testing, 116

S

Safe Deployment Practices (SDP), 39, 41, 45
Security, 1, 9
Security enforcement mechanisms, 55
Security Information and Event Management (SIEM), 200
Self-correcting digital nervous system, 167
Serverless infrastructure and event-driven meshes, 265
Service Bus, 237
Shared services, 83
Snapshot-based backup policies, 89
Software Defined Network (SDN), 76
Spatial computing, 263, 264
Spot Virtual Machines, 172
SQL Database, 4
Standard Azure Load Balancer, 97
Standard Load Balancer, 95
Storage Service Encryption (SSE), 189
Subnets, 49
Sustainable and carbon-aware architecture, 263

INDEX

T

Terraform, 115, 223
    CI/CD pipelines, 25, 26
    HashiCorp, 22
    modules, 25
    state management/remote back
        ends, 24
    workflow, 22, 23
360-degree visibility model, 158
Time-based retention (TBR), 189
Transparent Data Encryption (TDE), 189
Trusted execution environments
    (TEEs), 262

U

User-Defined Routes (UDRs), 56, 82, 133, 156

V

Velero, 118
Vertical Pod Autoscaler (VPA), 139
Virtual Machine Scale Sets (VMSS), 90, 118, 126, 146, 237
Virtual Machines (VMs), 3, 4, 178

Virtual Network Gateway, 83
Virtual networks (VNets), 4, 6, 9, 48, 80, 87, 128, 130, 153, 199
VM Scale Sets (VMSS), 4
VNet Peering, 49
Volume Shadow Copy Service (VSS), 103

W, X, Y

Web Application Firewall (WAF), 6, 10, 59, 61, 81, 87, 111, 221, 243
What-if, 31
Windows Subsystem for Linux (WSL), 205

Z

Zero Trust, 47
Zero Trust principles, 60, 80, 130, 196, 227
Zero-trust security model, 63
    assume breach, 68, 69
    conditional access, 71, 72
    data security/labeling, 70
    perimeter-based security, 66
    principle of least privilege, 68
    verify explicitly, 67
Zone-Redundant Storage (ZRS), 10

GPSR Compliance

The European Union's (EU) General Product Safety Regulation (GPSR) is a set of rules that requires consumer products to be safe and our obligations to ensure this.

If you have any concerns about our products, you can contact us on

ProductSafety@springernature.com

In case Publisher is established outside the EU, the EU authorized representative is:

Springer Nature Customer Service Center GmbH
Europaplatz 3
69115 Heidelberg, Germany

www.ingramcontent.com/pod-product-compliance
Lightning Source LLC
LaVergne TN
LVHW081347060526
838201LV00050B/1741